Conducting International Research and Service Collaborations

Conducting International Research and Service Collaborations: Tips, Threats, and Triumphs provides academic researchers, as well as non-profit and private professionals, with much-needed guidance on how to plan, implement, and manage international research and intervention projects.

Accessibly written and illustrated throughout with examples and case studies of projects from Robert B. McCall and Christina J. Groark's wide-ranging and decades-long experience of cross-border collaborations, this book outlines how to prepare for and ensure success of cross-border research projects and interventions, how to embrace unique circumstances you may encounter, and what to do if things go wrong. Each chapter covers a general domain of concerns, advice, and lessons learned in conducting international collaborative projects followed by concrete illustrations that pertain to them. Key topics covered include launching projects and working with stakeholders, travelling and living abroad, cultural considerations, planning and funding, administrative issues, dealing with crises, and successfully implementing and disseminating findings effectively.

This comprehensive guide is ideal for researchers and project managers – from large, global organizations to small NGOs, human services, private industry, and other fields embarking on such projects, as well as university students and academics.

Robert B. McCall, Ph.D., is Professor Emeritus of Psychology at the University of Pittsburgh. He is a former Co-Director of the University's Office of Child Development, an interdisciplinary unit devoted to university–community collaborative research and service demonstration projects. McCall is experienced in infant and child development and program evaluation.

Christina J. Groark, Ph.D., is Associate Professor Emerita of Education at the University of Pittsburgh. She is a former Co-Director of the University's Office of Child Development, an interdisciplinary unit devoted to university–community collaborative research and service demonstration projects. Groark is a specialist in children with disabilities and creating and managing domestic and international intervention programs for vulnerable children.

Conducting International Research and Service Collaborations

Tips, Threats, and Triumphs

Robert B. McCall
and Christina J. Groark

Routledge
Taylor & Francis Group

NEW YORK AND LONDON

First published 2022
by Routledge
605 Third Avenue, New York, NY 10158

and by Routledge
2 Park Square, Milton Park, Abingdon, Oxon, OX14 4RN

Routledge is an imprint of the Taylor & Francis Group, an informa business.

Library of Congress Control Number: 2021936942

ISBN: 978-1-032-02462-2 (hbk)
ISBN: 978-0-367-62787-4 (pbk)
ISBN: 978-1-003-18343-3 (ebk)

Typeset in Bembo
by MPS Limited, Dehradun

To:

The countless vulnerable children who grow up in unfortunate and sometimes wretched circumstances around the world, with hopes for improved conditions and families to nurture better futures for them;

Our colleagues at home and abroad, without their skills and dedication none of this would be possible;

Our spouses and children, who tolerated our long and frequent absences and supported our efforts on behalf of children.

Contents

PART III
Issues, Crises, Threats

Implementing an intervention or service will likely be
conducted by local colleagues, perhaps within an existing
facility. These people have various attitudes, policies, and
habits that may enhance or conflict with the quality of
the intervention

Sometimes a project needs to be started, but circumstances
make it very difficult to implement. If possible, it's best
to recognize these inhibiting circumstances as early as
possible, but you may simply have to do the best you can.
Other circumstances cannot be predicted or controlled,
such as the COVID-19 pandemic, so remain flexible

There can be misunderstandings that do not come to light
until much has already been completed. Try to reach some
sort of compromise

It's best to have discussions early about publications and
authorships, but sometimes the project must be completed
before authorship can be determined. Even then, a variety
of factors can complicate the issue

Applied projects conducted in community settings often
depend on the support of a variety of people, including
politicians, administrators, staff, and others, and any one of
these people can threaten to close the project down. It can
take a full-throttle effort to quell the attack

Relationships with your local collaborators are crucial to the
success of the project, and they may take continuous
nurturing throughout the project. Ruptures can threaten
the entire enterprise

PART IV
Conclusions

17 Benefits of International Collaborations

There are a variety of possible benefits to all the partners and
countries involved in an international collaboration

1 Why This Book?

The Need for International Collaborations

Collaborations are a key operational component of a great variety of projects, whether in private industry, non-profits, or academics. Many products, services, and research enterprises are multifaceted, each element requiring its own specific expertise. Most contemporary enterprises are just too complex for a single person or specialized team to complete the project on their own. Major manufacturing, such as building an airplane, requires numerous teams of designers, materials specialists, aeronautical engineers, and others working together to produce the final airplane. Similarly, the prevention of child abuse has economic, psychological, sociological, social work, medical, public administration and policy, and other facets. Even within academics, whereas single authored papers were common a few decades ago, now they are a rarity; collaborations among several specialists are the norm.

Further, for many purposes, a comprehensive approach involving several different perspectives is often found to be the most effective at achieving project goals. As someone once said, "If you want a product fast, go it alone; if you want a better product, collaborate."

Add to this trend the need for international collaborations. Manufacturing of a single product often involves parts created in numerous countries and assembled elsewhere. Academic publications are now often authored by an international set of specialists.

Behavioral research, for example, especially needs international collaborations. For many decades, the United States invested heavily in behavioral research, much more than any other country. An enormous literature was created, but it was largely focused on the subject populations of convenience. At first it was pigeons and the white rat, then college students, and now children who grow up in poverty, who are largely African American and Hispanic. But environmental factors play an outsized role in influencing behavior, much more than such factors influence atoms and electrons, for example. Consequently, much of the world's knowledge about behavior is based upon selected US populations, not even random or representative samples of American adults or children. Given the potential influence of a variety of environmental factors on behavior, many of which are not represented in such subpopulations, US psychology, for example, if not much of the world's psychology, is very narrow and of uncertain generality to the broader US population and certainly to the populations of other countries, especially where the environments may be very different. International collaborations could help to minimize this bias.

At the same time, the United States' higher academics are the envy of the world. Indeed, foreign student attendance at American universities, especially graduate schools, has been a major American "export" rivaling the export dollar value of major parts of agriculture and manufacturing. The investment the United States and increasingly other countries have made in research constitutes a "world resource" that can potentially contribute to alleviating much suffering and improving living conditions and human welfare in many parts of the world.

To accomplish the latter goal, international collaborations can be very helpful, if not necessary. Knowledge and skills from experienced countries may be crucial, because the status of these commodities in some countries is limited and sometimes outdated. But such knowledge and skills cannot simply be dropped into another country without being blended with the local environment, culture, history, policies, traditions, economy, academics, and religion. American projects, for example, that are simply "transplanted" or "parachuted" into another country often do not work very well. They can also be a source of social and political antagonisms. Other countries do not like to be perceived as inferior, and many are not, given the circumstances that exist there. Further, whenever a project seeks to improve something, the project itself implicitly declares that the status quo in that country is inferior or is not as good as it could be. Even if it is true and even if people recognize it, they don't like to be told implicitly that what they have is not very good. Yet, that may be the very reason they want to collaborate with Americans and specialists from other countries. It's a two-edged sword.

International collaborations are also good for Western professionals. American business, for example, recognized years ago that other countries may have a much less expensive workforce for manufacturing, some countries have more skilled employees than in the United States, and many countries represent a vast potential market for products that can be produced less expensively in those countries.

We can also gain insight into new parameters of behaviors, ideas on how to improve our own programs, and access to situations, populations, and perspectives that cannot be found in our own country. Americans, for example, rather commonly assume their approach is best, but that is not always so, especially when the context is quite different. Other countries have often created services, for example, that seem odd and perhaps ineffective to us but are well matched to the policies, economy, and social-cultural circumstances of that location. We do not always have all the answers for everyone. Usually, some shared learning with colleagues in other countries is required.

In short, there is great potential value in international collaborations for both Americans and the citizens of many other countries, but such collaborations require specialized skills of at least two kinds – knowledge of how to operate and function within collaborations and partnerships in general and how to function within an international context generally and specifically in the country in which the project is to be enacted. There is much more involved in an international collaboration than conducting the same project in your home country.

In 2006, the US National Committee for Psychology convened a group of psychologists who had substantial experience conducting international research projects to collectively identify the benefits and challenges of doing such work. While they cited several benefits to these projects, some of which are mentioned above, they recognized that "collaborations with researchers in other parts of the world entail moving back and forth across many boundaries, not only national, but cultural, linguistic, disciplinary, institutional, political, etc. Academic disciplines, for example, do not correspond exactly in different parts of the world....any field may be more theoretical or empirical in one setting or another....Ways of handling and managing data, including issues of ownership of and access to datasets, will not be consistent....Negotiations of power and status – revealed for example through expectations of authorship or control of research design, may be more complex. Conventional work habits including pacing, workloads, vacations, or sensitivity to deadlines and reporting requirements may vary. What is considered adequate protection for human subjects....and the concept of consent...is different" (U. S. National Committee for Psychology, 2006). Even the willingness to communicate or criticize can be very different.

We encountered all these issues and more, and that is what this book is about. In the chapters that follow, we describe in general and very concrete terms these and other potential problems that readers engaged in international collaborations may face that they would not likely encounter if the project were being conducted at home.

Purpose and Audience

This book relates the lessons learned by two American academics, a psychologist and a special educator, who conducted international collaborations over the past 25 years in countries on four continents pertaining to the rearing, development, and education of vulnerable children, many living in alternative care situations. This is not as narrow a focus as it may seem. Vulnerable children and those reared or educated in alternative situations in much of the world experience environments of extreme neglect, and neglect in one form or another is the most common form of child maltreatment worldwide. Further, while our projects focused on the development and education of such children, the lessons we learned about conducting international collaborations are much more general than the scientific focus or the contexts of our projects. This book is not about the *results* of our research projects in these countries; it is about the *process* of conducting these projects. Most of what is related in the pages that follow could be experienced in one form or another while performing nearly any kind of collaborative international project.

Although there are now a few articles available on resources for conducting collaborative and international work in general (Proctor, Vu, & Klonoff, 2019), there was nothing when we started in the early 1990s. We learned most everything by trial and error. We had some successes and a few failures, mistakes, and near disasters. We want to help others conducting such projects avoid some of these problems. Much of what we and others have learned has not been communicated – that is the purpose of this book. Unfortunately, the summary of the US National Committee for Psychology workshop cited above was never published. What we present in these pages does not appear in the scholarly reports of projects or even in articles aimed at colleagues who want to conduct international studies. We seek to provide individuals, whether academic researchers and scholars, students, non-profit professionals, or even private for-profit companies who may be contemplating, beginning, or in the midst of such a project, with the benefits of our experiences, guidance on what to prepare for, and illustrations of what can happen during such projects. We hope these tips help to smooth the way for such individuals and projects.

General Approach

Each project, country, set of participants, funders, specific partners, and other circumstances is unique, and these characteristics can change over time. So, we present general topical themes, guidance, and advice plus very concrete illustrations of why that advice may be needed.

Specifically, each chapter covers a general domain of concern in conducting international collaborations. The set of issues is not exhaustive, but it contains those concerns that we faced and that we regard as essential components of such projects. Each chapter begins with *General Lessons Learned* and some advice. These tips can apply to anyone about to engage in an international collaboration, whether an academic, service professional, or someone engaged in business.

This is followed by concrete *Illustrations* that pertain to those lessons. Some are deadly serious and professionally and personally threatening; some cases are triumphs or failures, either caused deliberately or by accident; and others are much less intense, even comical. Some represent events the reader may well experience in their own project. Others are more idiosyncratic, some marginally outdated, and they would likely never be experienced in the same way, but they illustrate the range of issues that can arise. They are included to create in the reader a broad set of expectations that embrace the variety of unique circumstances they actually may encounter.

Our Backgrounds

Although the lessons learned are quite general and applicable to a wide range of projects, it will help the reader to know our backgrounds and very briefly about the projects that form the basis of this advice and essentially all of the experiences.

One author, Christina J. Groark ("Chris"), has her doctorate in special education with a minor in statistics from the University of Pittsburgh. She had a wealth of administrative experience managing community service agencies for children, especially those with disabilities, before becoming an Associate Professor of Education and Co-Director of the University of Pittsburgh Office of Child Development (OCD), a position she held for nearly three decades.

The other author, Robert B. McCall ("Bob"), has a doctorate in experimental psychology with a minor in psychometrics/statistics from the University of Illinois. He had a long research career studying attention, perception, and memory in infants and general mental development in children birth through age 18 years. He also communicated research on children to applied professionals, policymakers, and the general public through magazine and newspaper articles and television appearances and programs. Bob then became Professor of Psychology and Co-Director of OCD at the University of Pittsburgh.

The OCD is a unique university "center" devoted to creating, funding, implementing, managing, evaluating, and communicating university–community collaborative projects to academic, professional, and policy audiences (Groark & McCall, 2018). Most of these projects pertained to innovative services and supports for low-resource infants, children, and families and the evaluation of community-created and managed services for these groups. In the course of 33 years, OCD funded, managed, and evaluated scores of such innovative services, all conducted in a partnership manner with community agencies.

From the standpoint of the international work that began in 1992, Chris contributed substantial experience in operating partnerships with diverse professionals; service/intervention program creation, implementation, and management; training of applied professionals; and programming for children. Bob had a long history of research experience, especially longitudinal studies of the cognitive development of infants and young children; research design and program evaluation; statistics and data analysis; measurement of behavior in young children; and communication through several media. Both authors had ample experience obtaining and managing research and service demonstration funding, and publishing literally hundreds of articles, chapters, and books. All of these experiences were relevant to the international projects that inform the ideas presented in this book.

The International Projects

Over the last 25 years, we have participated in a variety of international projects revolving around vulnerable children, some raised in alternative care environments, in various countries. Vulnerable children include those reared in extreme poverty or other adverse circumstances, those without well-functioning parents, or children and youth who are unmanageable for various reasons. Alternative care environments include caregiving services and facilities provided by adults other than the children's biological parents. These can include childcare for children living in extreme poverty, child and family services for families in extreme poverty, and residential care.

These projects involved creating and evaluating interventions to improve caregiving and education, assessing the development of these children both while in care and after they were placed into adoptive or other kinds of families, evaluating interventions created and operated by other organizations, training professionals to support families who want to foster or adopt children, training and certifying the preparation of US parents who intended to adopt children from foreign countries, assisting administrators in fundraising, and guiding academics in successfully publishing their work.

This is our experience, but we believe the lessons learned and described in this book can apply to a variety of professionals operating diverse projects in a host of different countries. Obviously, some countries are so similar economically, culturally, and academically (e.g., Canada, Great Britain, Western Europe, Australia) to the United States that much of the material we present on adjustment to the local country and its professionals will be largely superfluous. Conversely, we have not worked in remote regions of substantially underdeveloped countries, so there would be even more to learn about such enterprises than is offered here, although the general themes are likely to prevail.

The Illustrations in This Book

A substantial part of this book consists of concrete illustrations that provide examples of the general lessons presented in each chapter. These are written as experiences or as dialogues. We felt such dialogues would make the events much more concrete, alive, personal, and frankly interesting, than simple narrative descriptions. Also, this is the way the reader is likely to encounter these situations – not in abstract general terms but in concrete situations and dialogue with international colleagues.

Of course, some of these illustrations are inspired to some extent by our experiences, but they are not descriptions of actual events. Some are fictionalized. Individuals mentioned in illustrations are not real people. We recognize that this is not a common approach in academic writing, but it is common in other forms of communications, and this is not a traditional academic book. Basically, then, think of the illustrations as "docudrama," not history, that illustrate important points in the chapter.

Finally, this book, especially some of the illustrations, is written from our personal perspective, which inevitably represents an "American" point of view. Both of us were born, raised, educated, and professionally experienced in the United States. All of our characteristics are a blend of many factors, some of which are indigenous to the United States. It is nearly impossible and futile for us, as well as others, to try to identify which of our perceptions and interpretations are "American." We have seen foreign partners behave in ways that we could easily attribute to stereotypes of their culture, only to have other partners from the same culture behave quite differently. Such attributions are imprecise at best, often gross overgeneralizations, and usually not helpful.

Nevertheless, we acknowledge that this book is written from our perspective, both personal and American. We cannot escape it. We apologize for any insensitive statements; they were not intentional.

References

Groark, C. J., & McCall, R. B. (2018). Lessons learned from 30 years of a university-community engagement center. *Journal of Higher Education Outreach and Engagement*, 22, 7–29.

Proctor, R. W., Vu, K.-P. L., & Klonoff, E. A., Eds. (2019). Special issue: Multidisciplinary research teams: Psychologists helping to solve real-world problems. *American Psychologist*, 74, 271–406. Doi: 10.1037/amp0000458

U. S. National Committee for Psychology. (2006, October 5–6). International collaborations in behavioral and social science research. Unpublished manuscript summarizing a workshop held at Northwestern University, Evanston, IL, author.

Part I

Getting Started

Initiating, planning, and funding an international collaboration will take much more time, effort, and money than doing the same activity in your home country. We provide some guidance and advice on getting started in the chapters that follow.

2 How Projects Start

General Lessons Learned

Projects can start in many different ways. Some start in a very deliberate, planned, methodical manner, whereas others begin almost by accident as a consequence of fortuitous circumstances. We describe a continuum of circumstances and strategies that could occur. *First, in some cases, you may have an intervention or service program that has been developed and documented to be effective in your country, and you and/or an international colleague want to implement it in another country.* This comes close to a "transplanted intervention," but such projects are often not successful unless they are thoroughly matched to the circumstances of the host country with the help of local partners. Indeed, this is true even within a country, when an "evidence-based program" is to be "replicated" in a new location. Often the program implementers, the clientele, and a variety of other circumstances are different in the new than in the original location. The basic elements of such a program that are thought to be crucial for its success must be retained and are not negotiable, but other less essential facets of the program may need to be changed to fit the new local circumstances. Perfect replication of interventions and services conducted in community settings is a myth; there are always necessary modifications.

A second strategy is when an international site recognizes it has a problem or a need, and one or more interventions have been tried in your country or elsewhere to solve or reduce the problem. Then, a collaboration with partners from the host country is established, and together you work to create a new intervention that blends principles and promising practices learned previously with local circumstances. This differs from the first strategy in the amount of collaborative creation of a more-or-less new intervention, rather than tinkering around the edges of a previously documented program.

A third approach occurs when colleagues at an international location may have identified a local need and designed an intervention to address it, but they desire some additional expertise to fully implement and evaluate the effects of the intervention. Perhaps the originating party is a funder or a service organization that has a specific program in mind and wants an external and independent evaluation of the project they have created. Expertise in program evaluation, including design, measurements, and data analyses, may not be readily available in the host country, so the local professional or funder looks for such skills elsewhere. This situation requires a careful discussion of the roles, rights, and responsibilities of each partner and then continued adherence to those decisions. This can be challenging, especially when the evaluator has more experience with this kind of intervention, not just its evaluation, than the funder or local program creator.

One or another of the above strategies may actually begin in a rather unplanned manner. Perhaps you publish a paper or present at an international conference and a potential host country professional reads the paper or hears the presentation and it cues a particular need for program advice, evaluation services, or other skills. Contact is made, and a collaboration is born. Even more fortuitous, you may go to an international destination to provide insight and consultation about services, policies, or other activities. In the course of this consultation, a need for an intervention or other program emerges, and a project is inaugurated that no one had anticipated. At other times, an explosive event occurs in a country, and key players come to you for assistance. A new project is born out of need.

There are some general lessons that might be helpful.

First, be as collaborative with and accommodative to the needs of your international partners and situation as possible. Take time to learn as much as you can about relevant local history, operational procedures and rationales, skills, and so forth before providing too much advice. Spend time and effort on relationship building, socially as well as professionally, with your partners.

Second, at some point fairly early in planning, determine and be clear about your role and the roles of your partners and each's rights and responsibilities. Are you more-or-less equal partners in creating the project, or is your role only as an evaluator and not as a program contributor?

Third, be aware that funding international work is more difficult than a strictly domestic project. Planning such a project often involves more time and expense than a domestic project. It is very difficult to obtain planning funds to do the kind of ground work described above plus the time and expense of writing a grant application, getting ethical reviews, and so forth, all of which may require several international trips and be more complicated than usual. Some money from your university, employer, or other source is likely necessary to get started.

Fourth, grant money to fund the actual international project is also very difficult to obtain. If the project represents serious research on a major scientific issue, the US and Canadian governments and the European Union are able to provide the largest amounts of money, but special justification of a foreign location is required. If the purpose is more humanitarian in nature, there are some government agencies (e.g., USAID), international organizations (UNICEF, World Food Program, CARE), and foundations that will fund such projects, but the amounts of money tend to be smaller. Finally, there are private or religious non-government organizations (NGOs) that fund humanitarian causes, but these funds tend to be smaller yet. Just be warned, funding international projects, especially the planning of them, is difficult.

Fifth, pay attention to international politics, even though they may appear irrelevant at the time. For example, become acquainted with the economic and political health of the country in which you will work and the past and current relationship it has with your home country. Political events also can happen precipitously and change vital circumstances, impinging on your project and requiring non-traditional methods of research and service as well as substantial flexibility on your part.

Illustrations

"Want to Go to Ukraine?"

Sometimes a project does not start as a project. Suppose something like the following happens.

In 1992, Letitia, the Human Services Coordinator employed by the University of Pittsburgh Office of Child Development, attended a gathering at the university's Center for International Studies. There she talked with Katrina, the Director of a local agency that placed children from Eastern Europe in American adoptive families. Katrina was born in Russia, and she talked about the recent dissolution of the Soviet Union and the economic crisis that followed. As part of the fee for adoptive parents, she collected 10% for humanitarian purposes, and the former Soviet Union was certainly in need of humanitarian assistance at the moment. She had some general ideas about how to use the humanitarian funds and shared those thoughts with Letitia.

Letitia hurried back to the Office and burst into Chris's office. She took a deep breath and gathered up all the plastic casualness she could muster. "You want to go to Ukraine?"

"Sure," Chris knee-jerked, then paused. "Ukraine? What for?"

"I talked with this Director of a local agency that places children from Eastern Europe into American families. The Soviet Union has collapsed, its economy and that of its allied states has tanked, and people and children are desperate. She wants someone from the university with expertise on children to go to Ukraine and perhaps other Soviet countries to encourage services for children because they are being severely neglected."

"How are WE going to solve that need?" Chris muttered incredulously.

"That's what we go to find out."

"Mmmmm," murmured Chris. "Oooohhh-Kaaay. I need to sleep on this one."

Chris didn't sleep much. An assignment that vague sets the mind spinning through all sorts of possibilities. Helping the former Soviet Union deal with its most neglected children struck Chris's humanitarian chord, to say nothing of the intrigue of that part of the world and the whole amorphous idea. On the one hand, "Who are WE?" she thought. Then, "Why not US? This could be very interesting."

Some months later, Chris and Letitia went to the former Soviet Union. The local representative of the Pittsburgh agency arranged meetings with various political officials, administrators, homeless and runaway shelter directors, university faculty, and orphanage directors. Most of these people seemed overwhelmed by current conditions and the demands on them. But the directors of services for children and a few faculty members were the most interested in trying to collaborate to improve the children's well-being.

A year or so later, Chris and Letitia returned to the Soviet Union, and this time they took Bob along for another perspective. Most of the political figures they had met on their first trip were no longer in those positions. They were too much of a moving target. But the services directors and faculty were the same, and they wanted to continue to explore what might be done to improve the lives of vulnerable children. Bob was interested in getting the faculty members' ideas and blending research knowledge on children's development with action steps that might be pursued.

This began to feel like something big – who knows what might happen? Indeed, although it started as a fact-finding and consulting exploration, many years later after several additional trips and intense project planning, a large and comprehensive intervention in a traditional service organization for young children was launched. It took time, some money, patience, creativity, and a little bit of luck.

"She Needs an Independent Evaluation"

Sometimes your professional experience may be quite unusual. Other people hear about it who need that knowledge and those skills, and you get invited to perform a specific function in a collaboration. Suppose several years after conducting and evaluating international interventions to improve the development of vulnerable young children the following occurred.

Chris and Bob had just finished making a presentation at a major national conference of child development researchers on one of their international intervention projects. There are only a few people in the world that we knew of who did this kind of work. The project was still ongoing, but we described the intervention and presented some preliminary results.

"Hello, Bob and Chris. My name is Paul, and I work for a large training institute on the West Coast," said a fairly tall man who had just attended their presentation. "Do you have a minute?"

"Sure," replied Bob. "Can we go off to those chairs on the side where it is quieter?"

"Great" agreed Paul. After sitting down, he continued. "The project you just described is very interesting, and it is similar to a project that my colleagues and I have recently become involved in.

"We've been contacted by a young woman who has just started a new foundation, and she is interested in improving vulnerable children's development. She has traveled around the world visiting services for infants and young children, especially in Eastern Europe, Africa, and Latin America. She has some ideas on how she wants to improve these services that are similar to what you have done. She wants us to train the caregivers and service personnel, because we are the largest trainer of childcare staff and teachers in the United States.

"She is a very vivacious, energetic person – no moss grows under her feet. She is a go getter – things are going to happen at her hand. She has no qualms about scheduling an appointment with the President of a country and telling him how she could contribute to improving his services for children. It helps, of course, that she has her own money and is not asking the President to pay for it."

"That's fascinating," said Chris. "How do we relate to this?

"We have told her," continued Paul, "that if she wants to improve more than one institution, she needs to have an evaluation conducted on her first intervention so she can document to political figures the benefits of her intervention. She intends to raise money from other people, governments, the public, whomever to do this on a big scale. She will need some evidence that comes from a credible source. We have done some evaluations – we could do it – but we are doing the training of the service personnel, so we are not independent. She needs an independent evaluation. Would you be interested in doing that; can I suggest she consider you for that role?"

"Yes, certainly," replied Chris. "Don't you think, Bob?"

"Sure," agreed Bob. "We are happy to have you explore this with her. We operate somewhat similarly to your unit in that we need funds to do almost anything."

"Of course," said Paul. "She hires us to train, and I am sure she would hire you to do the evaluation if that worked out. I'll talk it over with our Director, and we'll let you know if we can proceed to discuss this with her."

"Actually," Bob added, "we have an entire division within our Office of Child Development that is devoted to evaluating community-created and operated service programs of different kinds, many of which pertain to young children. So we are accustomed to planning evaluations with service professionals and are used to accommodating to applied circumstances."

"That's great," effused Paul. He went on to arrange a "get-to-know-you" meeting in a Latin American country for us to meet the Foundation Director and to get more information. After observing the services and meeting with local academics who believed they could train and supervise data collectors, we felt confident that a reasonably rigorous evaluation could be conducted. We then met with local stakeholders and conducted a short logic model session to be sure we all agreed on the same general goals and procedures. This exercise gave us the basis for designing a thorough evaluation of the intended intervention, which we then proposed to the foundation.

"I Have a Proposition for You"

Another situation might be similar to the previous example. Suppose the following occurred.

Chris and Bob first met Dr. James on a trip to another university to consult with faculty there who were operating a major project following up children adopted by families in the United States. While there we met Dr. James, who is the world's leading authority on the physical growth and medical condition of vulnerable children in a variety of countries, especially those who have been adopted into the United States. As a result, he was an advisor to several international organizations concerned with vulnerable children and those adopted to the United States.

In particular, Dr. James was on the Board of Directors of one of these organizations in China. Although it supported a variety of projects and services in China, their main program consisted of sending specially trained supplementary caregivers to service agencies for three hours on several days of the week to stimulate and play with the infants and toddlers there. That was an experience these children would otherwise not get.

A few years later, we met Dr. James again at a professional meeting, and after a long day, he invited us to have some refreshments with him.

"I have a proposition for you," he began. He described the foundation and their program of stimulation of the infants and toddlers in China, and that he has been on their Board of Directors for many years. "A little while ago, I did a simple and small evaluation to see if the program of stimulation improved the children's physical growth."

"Did it?" asked Bob.

"Yes, a little," Dr. James confirmed. "But the study was so small and the circumstances not the best, so it really does not constitute a serious evaluation. This program is really quite a large effort and operates in many locations in China, and I think they need a more comprehensive evaluation by an independent group, and you have done more of this kind of thing than anyone. As a Board member, I want to propose to them that they hire you two to conduct such an evaluation."

"That's very kind of you," responded Chris. "We appreciate that you have confidence in us to be able to do this. Actually, as you may know, we have a whole division of program evaluation within the Office of Child Development, and it is headed by a very bright Carnegie Mellon University graduate who happens to have been born in Shanghai. Moreover, he has adopted two Chinese girls, so I am sure he would be interested in being involved in such a project."

"And because he is Chinese and speaks Chinese," Bob added, "he could help immensely in implementing the evaluation and working with the agency administrators and caregivers."

"All the better," confirmed Dr. James. "I have a Board meeting in a few weeks in China. Let me explore this idea with them in principle. If they agree, then we can be more specific about what could be done, where, and how extensively."

A few weeks later, we were invited to meet with the foundation's representative in China at our expense. Once again, professional contacts led to an unexpected opportunity, and having some travel money was necessary to be able to meet and become acquainted with the foundation's representative and to see the intervention in operation. Also, the fortuitous coincidence of having our own staff member who is Chinese may have contributed to getting hired.

Having an Intermediary

The above three illustrations all involved a professional intermediary who suggested us to a funding agency to evaluate their program. Consider the following situation in which the intermediary is not a professional colleague and the project is not decided yet.

A friend of Chris and Bob met in Washington, DC, with a diplomat from a Western Asian country. He was acquainted with an entrepreneurial group of talented women from wealthy backgrounds who had established a non-governmental organization to assist women and children in their country. They had an interest in helping vulnerable children, providing specialized therapeutic services to children with disabilities, and helping adoptive parents cope with some of the adjustment problems that some adopted children present.

Our friend encouraged us to visit this group to learn first-hand what they wanted to do and to share our knowledge with them. After years of experience with such startups domestically and internationally, we knew we needed to be on-site to fully understand the current situation and its relation to the goals of this group. Thus, once again, the first such meeting was initially self-funded with no guarantee of a future project and funding. The visit provided us with vital information about what could be done to help them achieve their goals, which were not yet firmly established, and it gave our hosts confidence that we were appropriate for their needs and sincerely interested in working with them over time.

Policymakers Giveth and Taketh Away Projects

Sometimes an international project can take place in your home country, and political events beyond your control can initiate, modify, and end the project in almost a continuous stream of changing events all transpiring within a few months. This was a very unusual situation, but it can happen. So, expect the unexpected in international work.

The unexpected birth of a project. In the summer of 2012, Vladimir Putin, President of the Russian Federation, suddenly instituted legislation that restricted American families from adopting Russian children unless those parents could document that they had successfully completed a training program that included 13 topics specified by a group in Russia. These topics ranged from basic child development issues to details of Russian adoption legislation.

This restriction was levied immediately on all potential adoptive parents regardless of what stage of adoption they were in. For instance, some parents had already met their child several times and had bonded in a variety of ways. Others were prepared with all legal documents, including travel itineraries, and had set appointments to attend court proceedings in Russia to finalize their adoption.

It was a crushing blow to US adoption agencies and especially prospective adoptive parents. It was a policy without an implementation plan. There was no Russian or US entity that provided the required training in those topics, and this was all new to parents, adoption agencies, and trainers who prepared adoptive parents in the United States.

The phone rang early one morning in Chris's office. It was Katrina, the Director of the Pittsburgh-based adoption agency specializing in placing East European children. "Chris, are you sitting down? Terrible news has been announced in Moscow. Our American families who are prepared to adopt Russian children are in a holding pattern unless they can receive training through a Russian-approved institution in the next 30 days! Otherwise their documents will expire! We have many families in the United States, not just in Pittsburgh, facing this crisis immediately!! Some are already in flight on the way to Russia!"

"Whoa! That IS terrible," exclaimed Chris. "But why are you calling me? The university and the Office of Child Development are not Russian-approved institutions for anything! How can we help?"

"I'm coming right over to talk with you. We have got to strategize something immediately."

In the next few days, Chris organized a team of university faculty, social service professionals, and medical agency representatives from Pittsburgh and across the United States who agreed to contribute to a webinar that would cover all 13 essential topics required by the Putin administration. They were all happy to help. Then, within a few days, a few technical experts designed a platform to test the knowledge gained by parents who attended the webinar.

Further, the university had to quickly assemble a portfolio of certifications and licensing documents from US entities to attest that OCD was a legitimate part of the School of Education, that the School of Education was a legal unit of the University of Pittsburgh, and even that the university was qualified to create and deliver training in each of the 13 modules! Then, documents had to be created and assembled in a portfolio, and each document had to be notarized, signed, and stamped with an institutional emblem and sent overnight to Moscow for review.

Many favors were called in a few days. A network of colleagues was assembled, the curriculum described, and the documents produced. The entire portfolio was approved by the university and the Russian government. The webinar was held for parents from across the United States who were in the final stages of adoption, some of whom were already in Russia waiting to pick up their children.

However, the project was not yet fully developed and implemented. The Russian approval of the training curriculum stipulated that each of the almost 50 oblasts/regions of the Russian Federation had a central court judge who would have the authority to dismiss/deny any document presented by a family desiring to adopt a child in that district and could request additional documentation of the parents' completion of the required training.

And many judges did just that. Our staff had to respond at all hours of the night due to the time differences (Russia is at least seven hours ahead of the United States and has nine different time zones within the country) to create new documentation at the request of any judge presiding over the dozens of adoption cases in progress. Each adoption file had to contain documents attesting to the authenticity of the training and testing results, and each document had to be notarized and signed personally by Chris.

The saga continued and worsened. Within 30 days of the start of this parental training program through webinars, this method was deemed unacceptable by authorities in Moscow: in-person training was required. Once again ingenuity and flexibility prevailed. Weekly training sessions of professionals in the adoption field were provided in Pittsburgh. Professional attendees came from all over the country, from as far as California and Texas to the smaller counties in Pennsylvania near Pittsburgh. Then, these professionals in turn went back to their home locations and provided the required face-to-face training of adoptive parents. Parents had to be tested, and the University of Pittsburgh had to certify – and Chris had to sign the documents attesting – that they had successfully completed the course. As you can see, this program was labor-intensive, and required nontraditional efforts, methods, and work hours.

The death of the project. Almost as soon as this project was created, it was killed. Four months after it began, heart break slammed all remaining participants in this program. In December 2012, Putin signed a bill to ban all Russian adoptions into the United States. The training and certification program was ended; potential families were left separated forever.

International politics can giveth and taketh away projects without warning. Expect the unexpected; monitor international politics; and be prepared, creative, and flexible.

3 Study Some History

General Lessons Learned

It's important to have some knowledge of the history of the country, especially events and conditions that might influence the people and organizations with which you will be working. This can include major political events and armed conflicts, economic conditions, the history and traditions of the institutions that you will be working with, and the country's societal values and priorities. Also, if you intend to visit a country several times and stay a week or two on each visit or if a colleague will live in the country of your project, then becoming familiar with that country's history becomes even more relevant. It can help you understand customs and the behaviors of individuals and be better able to relate to local professionals and citizens.

Illustrations

International projects potentially could be rooted in any country of the world. To illustrate the need for a knowledge of history, we have selected three large countries and areas of the world as examples.

Soviet and Russian Federation History

Suppose you intend to work in the Russian Federation or any of the former Soviet States, presumably on services for vulnerable children. Then, you need to appreciate some aspects of Russian history. Although the former Soviet States have their own individual histories, their services for children are still quite similar and stem from their previous association with Russia.

Russian political, social, and economic history. We started by reading a serious academic history of Russia. Basically, the main events were the reign of the Tsars, the Bolshevik Revolution, and the fall of the Soviet Union. The major theme running underneath each of these eras was that the Russian people were essentially serfs of the government throughout history. They were experienced in having the Tsars exercise great power over them, and while the Bolshevik Revolution in 1917 promised some reforms, the people were accustomed to having the Soviet government dictate much for them. But this same disposition made the fall of the Soviet Union in 1991 especially crushing, because all that government framework and the social and economic support it provided suddenly disappeared, at least for several years.

By the 1980s, after the Bolshevik Revolution, the government had provided a framework that essentially propped up society. There was free medical care, a guaranteed job and salary, free housing and childcare, and cheap vodka. People did not have to worry about the future, because the state would provide for it. The quality and extent of these provisions can be debated, but there was some degree of personal social and economic security and stability for most people.

Also, military might was the coin of the world. A good deal of Russian industry was devoted to supporting the military, and indeed Russia was a world military power, rivaling and sometimes exceeding the United States on some dimensions. Further, despite his despotism, Stalin was the nation's hero according to some surveys, because "he defeated the Nazis in World War II." So, the people were accustomed to a strong and even repressive government but one that ultimately provided some degree of social and economic support and stability, and they also had pride in their nation as a world power.

Then, in 1991, the Soviet system collapsed, with dire economic consequences. The military-industrial complex dissolved, and with it many jobs. For example, the majority of manufacturing in St. Petersburg had been devoted to the military, but now most factories were vacant. A taxi driver might be a former nuclear physicist who could not find other employment. Food and housing were no longer guaranteed, and many people received meager pay or had no job at all to support themselves. Further, the world criterion for status had shifted somewhat from military might to economic prowess. Therefore, people were both socially and economically in despair, the future was uncertain, and their nation had lost some of its international status.

A major consequence was the deteriorating health and well-being of its citizens. Russian health, longevity, and mental health had never been good by international standards, but they got worse through the 1990s. An article in *The New York Times* in December 3, 2000 (Wines, 2000) painted a desperate picture. Life expectancy was dropping by the month to an average of 65.9 years, nearly 10 years less than in the United States, and this figure hides the dismal expectancy for men, which was only 59.9 years (it was 72.4 for women). Between 1990 and 2000, the death rate rose by one-third to the highest of any major nation, the birth rate dropped by almost 40% to nearly the lowest of any nation, and the rate of newly disabled people rose by 50%.

Life expectancy, of course, is an indicator of a cluster of maladies, including poverty, stress, poor nutrition and diet, and high rates of smoking and alcohol use. The Russian diet has always contained a good deal of fat (i.e., butter, sour cream), smoking is common, and vodka and moonshine are staples. The average Russian consumed 4.4 gallons of alcohol per year, the world's highest. It is part of the culture; when Gorbachev invoked a series of restrictions on alcohol, he was nearly run out of office. Indeed, it is commonly believed that nearly nothing happens in Russia without alcohol. A shot of vodka may be necessary in the morning before a repairman will start to work, or one may need to bring out the vodka to make much progress at a 9 am meeting with a major administrator.

The leading killers at the time were cardiovascular disease, violence, and accidents. Medical care and hospitals were short of money, drugs, and syringes, and the Soviet concept of free medical care basically disappeared. Rates of accidents, homicides, and suicides increased markedly.

In the 1990s, bread lines were common – if there was any bread at all. No one was in the stores, and the stores had little to sell. Eating at restaurants required calling in advance to determine if they had food to serve. You ordered in advance, and the owner perhaps had only one bottle of beer and had to go down the street to get more. Some "walls" in the airport were made of canvas, and everything looked old and in need of repair. Moreover, repairs seemed largely "make do" or "patchwork" in nature.

All of this was very relevant to any future work pertaining to the care of vulnerable children. The traditional method of caring for such children was to put them into institutions. Although birth rates declined during the 1990s, the institutions were full of children (see below). Prenatal diets were probably worse than usual, and prenatal alcohol use was higher than the already lofty rates. This likely led to higher percentages of infants with disabilities and fetal alcohol syndrome in the institutions, and these conditions are the ones that can be immediately recognized. What about the more subtle influences on children that are not apparent in infancy but that may become manifest in different ways years later? In fact, research has revealed that institutionalized children born in the late 1990s and adopted by advantaged American families had somewhat higher rates of behavioral and related problems later in childhood and adolescence than children born later in the 2000s when social and economic conditions had improved.

Social and economic conditions in Russia indeed improved very substantially over the next two decades. Today, there are large modern shopping malls with crowds of customers and stores bursting with merchandise, major supermarkets have been introduced, cars are prevalent, and St. Petersburg and Moscow are basically similar in these respects to most major cites of the world. Indeed, these are two of the most expensive cities in the world for visitors.

History of institutions for orphaned and vulnerable children. Institutions to care for orphaned and vulnerable children have a long historical tradition in many countries, including Russia. Tsar Fedor Alekseevich (1676–1682) provided public care in institutions for abandoned and unwanted children similar to those created by the monarchies of

Europe at the time. Peter the Great issued a decree in 1712 that established hospitals for the "children of shame" to be supported by the families of the Tsar and wealthy nobles.

Decades later under Empress Catherine II, in response to reports that some children were being abandoned and some died or were killed by their parents, Russia adopted a more humanitarian attitude by providing two large institutions (*doma*), one in Moscow (1764) and one in St. Petersburg (1770), that gave refuge, nurturance, and education to innocent children born to unwed mothers or parents too poor to rear them. Admission was quite liberal – nearly any infant or child was eligible. In addition, rural peasant women were paid to nurse infants and care for children, the seeds of foster care.

This approach lasted for more than a century. By the second half of the 19th century these institutions cared for 9,000–17,000 children, and 30,000–40,000 children were under some form of substitute care per year in these two cities, then funded by foundations and other private sources rather than the government.

These arrangements had shifted the emphasis toward humanitarian care for the children as opposed to the European concern for the welfare of the mother. But, in addition, in the late 1800s the role of the biological mother was emphasized in the fostering process. She was encouraged to feed and care for her child to promote attachment and the child's well-being, even if she could not otherwise fully care for the child.

By the time of the Bolshevik Revolution in 1917, this system became unmanageable. There were too many children, too few wet nurses and foster families, too much illness, and too many foster families who were more interested in the fee than in providing good care. So in 1918, the new Soviet government did away with the private institutions and fostering system and established a network of state-supported institutions that had a different philosophy. The government recognized the need for women to have babies, but everyone, including mothers, needed to work, so the state would help provide care for children whose parents could not raise them fully on their own.

Initially this consisted of "mother and child homes" established within the state's health system. But later, social and economic circumstances led to the creation formally in 1946 of "Baby Homes," institutions for infants and children up to age four whose parents could not rear them and eventually for children with disabilities. But older orphans were often portrayed in newspapers and books as pawns of criminals, which shifted public opinion toward viewing all such children as outcasts and undesirables who should be segregated from the rest of society.

After the fall of the Soviet system, formal and informal philosophical attitudes toward orphaned and vulnerable children moved toward fostering and supporting families to care for these children. But the very substantial social and economic devastation of the 1990s prevented any implementation of these attitudes, and the Baby Homes persisted in more-or-less their original form.

Historical legacies in contemporary Baby Homes. Baby Homes and other institutions are still the predominate form of care for vulnerable children, and several specific historical themes persisted through the 1990s and 2000s and largely continue today.

One theme is that the state-supported network of institutions was available to care for infants and toddlers whose parents could not, or would not, rear the child. Given the social and economic hardship of the 1990s, substantial numbers of infants were relinquished to the Baby Homes despite the lower birth rates. As a result, by 2004 there were 255 Baby Homes in the Russian Federation housing approximately 19,000 children birth to four years of age.

A second persistent theme is that it is relatively easy for parents to relinquish a child in Russia compared to the United States, for example. A parent only needs to write a letter to the relevant magistrate to give their child to the Baby Home. The most common reasons for relinquishment are 1) financial inability to rear the child, 2) parental inability to behaviorally rear the child (e.g., substance abuse, mental health issues, mental or physical incompetence including being a teenage mother), 3) parental unwillingness to rear a child with disabilities, and 4) involuntary termination of parental rights (i.e., abuse, neglect). Some children are given up permanently, others temporarily, but if parents do not visit the child in the Baby Home at least once in six months, parental rights can be terminated.

A third enduring theme is the pervasive parental and societal avoidance of children with disabilities. For example, in 1994, 44 children were born in St. Petersburg with Down Syndrome, and all but 2 were placed in the Baby Homes. The practice was so common that nurses in the obstetrics wards of hospitals reportedly would encourage parents to relinquish a child with disabilities immediately to the Baby Homes; the nurses made similar suggestions to teenage and single mothers of essentially typical newborns.

The Baby Homes are administered by the Ministry of Health, a fourth enduring theme, and each is directed by a pediatrician. Government figures on the number of children in the Baby Homes who have disabilities are difficult to interpret. This profession is focused on illness and disabilities. Further, Russian medical diagnostic categories include the equivalent of "failure to thrive" and "minimum brain dysfunction." Not only are most infants who arrive at the Baby Homes underweight and may have experienced a variety of adverse birth conditions, but the Baby Home environment does not promote typical development thereafter. The result is that the vast majority of infants and children in the Baby Homes are considered to have one or another medical conditions or disabilities by these definitions.

Another consequence of the continued administration of the Baby Homes by the Ministry of Health is that the staff receive some training in health care but no training in the social and emotional development of infants and children or how to handle their behavioral issues. Further, Russian society, similar to that of the United States several decades ago, does not embrace children or adults with disabilities and prefers to keep them out of public view.

A fifth theme is the form of early education. The Baby Homes do provide some instruction, but it is mostly focused on education for children with disabilities. Further, the style of education and the manner in which adults behave with young children was carried over from the Soviet preschools or "yasli-sads," which were also under the Ministry of Health. They used a very detailed curriculum and prescribed teaching methods that were extremely teacher-directed and controlled, and this style of interacting with preschoolers was carried over to the infants and young children in the Baby Homes.

Voices of change. As early as the 1980s, some professionals in the Moscow Psychological Institute championed the views of Lev Vygotsky, Maya Lisina, and John Bowlby, who emphasized that the child's early learning and development was sourced in the child's social and emotional relationship with an adult. Subsequently, researchers published a paper (Galiguzova et al., 1990) criticizing the way infants were cared for in the Baby Homes and urging smaller groups of children, more stable relations with individual adults, and mixing children of different ages. But nothing actually changed in the Baby Homes.

Similarly, with the fall of the Soviet system and its prescriptions for the care of infants and toddlers, other professionals and governmental groups began to advocate for changes in the care of children in the Baby Homes and other contexts. For example, in 1992 the city of St. Petersburg accepted a city-wide social project called "Infant Habilitation," which prescribed screening, assessments, and new services for infants with special needs and their families. An interdisciplinary team of specialists urged a family-centered, parent–infant interaction approach to promote the development of children birth to three years of age who had developmental delays or were at risk of delay.

To build upon this initiative, a new preschool was established, eventually called the Center for Inclusion, that integrated a few children with disabilities into their preschool program, which in addition to early education had an emphasis on the social, emotional, and personal development of the children. Once the Center was established and operating well, the leaders sought to transfer the same ideas to the Baby Homes. Many people, but not all, were supportive of these principles. Even Baby Home caregivers suggested on a questionnaire in 1998 that they could do a better job if they had fewer children in a group, the children had one or two caregivers who were always available to them, and children were not routinely moved to different groups of caregivers and peers. But no model or strategy for implementing these ideas emerged, only broad principles, so little actually changed in the Baby Homes.

Institutions for infants and toddlers in international perspective. It is important to perceive the Russian institutions for infants and toddlers in international perspective, because they are in broad strokes similar to institutions in many other countries.

A good strategy is to have a graduate student review the research and professional literature on the structure and organization of such institutions worldwide and the nature of the care they typically provided (Van IJzendoorn et al., 2011). This can be challenging in this case, because most articles provide only narrative reports, some second or third hand, of the institutions that provided children for international adoption. Reports that gave first-hand descriptions were likely more accurate, but they were only available from the institutions that agreed to work with the professionals, many of whom were foreigners. So the sample of institutions available in the literature is likely not representative of all institutions and not of all countries that had institutions. Nevertheless, it can provide general descriptions that matched quite well the nature of the institutions in Russia and many other countries.

Specifically, institutions vary from one to another between countries, within a country, and even over time within a country, so there is no one description of an institution for infants and young children. The professional literature largely consisted of studies of internationally adopted children, mostly placed into Western Europe and North America, from China, Southeast Asia, Latin America, and Eastern Europe, especially Russia. Nevertheless, with some notable exceptions, this literature revealed a few broad characteristics across many such institutions. These common themes were:

- Children are cared for in rather large groups in a single room, typically 9–16 children, although it could be much higher, as high as 51 infants in a single room. There are many children per caregiver during the daytime hours, approximately 8:1, but this too can be much higher.
- The groups have children of similar ages, perhaps separated into infants, crawlers, walkers and toddlers, and older children.
- Children are routinely "graduated" to new groups of caregivers and peers as they age.
- Children with disabilities are housed in separate rooms or even different institutions.
- An individual child typically will experience numerous and changing caregivers, even within a week but especially over months. Other adults may provide specialized services (e.g., music, physical activity) or be visitors (e.g., prospective adoptive parents, volunteers).
- Caregivers get minimum training, and that which they do receive focuses on care routines and health, not social-emotional and behavioral development.
- Caregivers spend most of their time providing basic care (e.g., feeding, bathing, changing, preparing food), often in silence, rather than in interaction with children (e.g., talking or playing with them).
- Caregivers perform their duties in a business-like, sometimes perfunctory, manner, with little eye contact, talking, or warm, sensitive, responsive interaction.
- The daily routine for children is often highly regimented; the activities of all children in the group are enacted according to a single time schedule.

Therefore, the general experience for children is impersonal, with numerous and changing caregivers and peers, and little individual attention or interaction with a warm, responsive, consistent adult. By most contemporary definitions, this constitutes "extreme neglect," and neglect is the most common form of maltreatment of children in the world.

Latin America

Latin America is another area of the world in which international projects might be conducted. The major countries of Central America, for example, formed an alliance in 1823 called the United Provinces of Central America, but this union dissolved 17 years later, and each country traveled its own path thereafter (https://en.wikipedia.org).

But their individual histories were each characterized by frequent changes of political leadership, some peaceful but often violent. El Salvador, for example, had massacres of nearly 30,000 people twice, once in 1932 and again between 1979–1981, and Nicaragua had a revolution in 1979.

It is important for US personnel to recognize that the United States was not necessarily an innocent or unbiased player in some of these social and political upheavals. Youth from El Salvador, for example, formed gangs in California, and the United States deported many of them, who became the "Maras" gangs in El Salvador. The US government also orchestrated a sustained campaign of embargoes and subversion in the changing political history of Nicaragua after its revolution of 1979. Both countries recently experienced a good deal of social and political unrest and violence, so it may be especially important for you and international colleagues to steer clear of political discussions.

Central America is also on the geological Ring of Fire, so it is quite prone to earthquakes and volcanoes, and it is often washed over by hurricanes and tropical storms. The region experienced Hurricane Mitch in 1998, the second most deadly Atlantic storm in history in which 11,000 Central Americans died. El Salvador, for example, had a massive earthquake that killed 1,200 people and rendered perhaps a million people homeless in 2001. The Santa Ana volcano erupted, followed by a tropical storm in 2005, and torrential rains hit in 2009 and 2011.

Between the history of political instability, social violence, and civil war plus natural disasters, international colleagues need to be a bit more aware of the social, political, and environmental surroundings than they might be in some other countries.

China

China is another country in which international collaborations often transpire. China has a 3,000-year written history dating from the Shang Dynasty (c. 1600–1046 BC) that is characterized as a succession of Dynasties. In 1911, the

Xinhai or Chinese Revolution overthrew the last Qing Dynasty, but this led to a fractured country with several different attempts to unify and control it.

Japan occupied the country from 1931 to 1945 and committed terrible atrocities. After the end of World War II, civil war broke out for control of the country between the Communist Party of China and the Nationalists, both of which also committed mass atrocities, killing millions of people. The Communists defeated the Nationalists, who then left the mainland for Taiwan.

Thus, the People's Republic of China was established on the mainland in 1949 under Mao Zedong. Over the next 27 years, Mao conducted campaigns and 5-year plans imposing massive social and economic collectivism on the country, including the "Great Leap Forward" and the "Cultural Revolution" in which an estimated 45 million people were killed by execution, forced labor, and other atrocities. After Mao's death in 1976, a power struggle ensued, and eventually Deng Xiaoping became the de factor leader of China from 1978 to 1992. He opened up the economy to a mixture of central control and free market activities that eventually raised 150 million peasants out of poverty. His leadership was not without uprisings, however, most notably marked by the "Tank Man" of Tiananmen Square in 1989.

As can been seen, China's history parallels that of Russia and in some ways Latin America in which the people essentially have been continually dominated by oppressively controlling governments and millions of them were killed in the process of evolving from monarchies or despots.

References

Galiguzova, L. N., Mesheriakova, S. J., & Tcaregorodtceva, L. M. (1990). Psychological aspects of children's upbringing in baby homes and children's homes. *Voprosi Psychologii*, *6*, 17–25. (in Russian)

Van IJzendoorn et al. (2011). Children in institutional care: Delayed development and resilience. *Monographs of the Society for Research in Child Development*, *76* (Serial No. 301), 8–30. Doi: 10.1111/j.1540-5834.2011.00626.x

Wines, M. (2000, December 3). An ailing Russia lives a tough life that's getting shorter. *The New York Times*, *1*, 118–119.

4 Know the Country and Its People

General Lessons Learned

It is very important to get to know the country you will work in, especially its history, culture, economics, and people. Read a travel guidebook to the country in which you will work. It can provide a great deal of practical information about sights, cultural landmarks, food, local transportation, weather, holidays, and so forth. Although it's advantageous to know as much as possible before you start, the real learning occurs over the entire time you work there. This knowledge helps you understand the country's economy and priorities and things your colleagues do and say; it also facilitates getting to know them as people, even personal friends.

A country's political and public image, often as conveyed by the government, may be quite different from the impression given by the local people. For example, our impression of some Eastern European politicians was that they were a bit blustery and not terribly friendly, but the local people were kind, caring, and personable. Many had next to nothing but would eagerly share it. It was officialdom you might be concerned with, not the local people.

Illustrations

What follows are some illustrations of things that could happen that collectively contribute to a picture of the contemporary culture, economics, and social life of the countries in which you might work. They illustrate the range of experiences you may have with the local people, and most were positive but not all. While you cannot prevent some unpleasantries, you can minimize them and simply accept them when they occur.

Russia

Chris's first impressions. It was the early 1990s, and I was on the last leg of my 20-hour first trip to Russia feeling exhausted, uncomfortable, hungry, and a little apprehensive. I couldn't sleep on the plane, and I spent most of those hours dwelling on thoughts that encompassed all the stereotypic visions I was fed as a child and young adult regarding communism and the despair, fear, and difficult living conditions it caused Russians. The recent economic decline likely made it all worse, and the fact that my mother had spent time in a Russian concentration camp after World War II didn't help my image of Russia. I wasn't sure I was ready to land, but I was also very sure I wanted the adventure to begin.

Through all my mixed emotions and wariness, I could feel we were descending. The confirming announcement by the flight crew only helped to heighten my anxiety and skepticism of having agreed to this trip. It was November. It was dark outside, and as we landed, all I could see was snow and ice draping everything at the airport. When we slowed to a stop, I could see men clad in what seemed like military uniforms waiting in the freezing temperatures for our arrival. They were unsmiling and held rifles. Not a welcoming image.

The plane stopped, and so did the chatter among the passengers. There was an eerie silence among 40–50 people. Over the next years I observed frequent instances of this lack of talking in public places. There were very few of us who did not speak Russian, so we followed the example of those who did and moved quietly and slowly to disembark the plane. We entered a rather small, crowded, dingy room. We just waited there, in silence. Then, suddenly the crowd moved almost as one mass; I followed the other passengers who seemed to know the routine.

The next holding area was overheated to compensate for the frigid temperatures outside and contained cubicles that housed official-looking passport control personnel. Over each small cubicle a red light signaled stop; it turned green when the next passenger was permitted to enter. No one smiled. No one talked. A rifle-toting soldier stood outside each cubicle. Quite intimidating, but maybe that was the intent.

I waited my turn in the queue with my heart pounding so hard I could feel it in my throat. My entrance papers were wilting from my clammy hands. The light turned green; it was my turn. I entered the cubicle.

The stern face of a rather large female agent in military-like uniform was dimly visible through the glass window of the eight-foot-high cubicle. In a rather loud, rough voice, she demanded something in Russian. I stood facing her, frozen in my spot. She repeated the demand. I stood motionless. Again, the booming, unamused voice came at me, and suddenly I realized she was saying "PAPERS" in English! I peeled all the papers I was holding out of my trembling right hand and unintentionally shoved them at her. She rifled through them and immediately pushed half of them back to me, retaining only two or three documents, including my passport.

The drama continued. Minutes, that felt like hours, went by in total silence. She looked me up and down. By now my entire body was shaking, and I was in a cold sweat, piled with winter scarves, gloves, and the heaviest coat I owned in an overheated cubicle with bright lights shining in my eyes.

Abruptly she grunted something in Russian and gestured with her hands for me to remove my glasses and step back up against the wall behind me. I guessed I was not wearing glasses in my passport picture. Suddenly, I heard banging that I recognized as official stamps being pounded into my passport. She shoved all my documents through the window to me without a word. Immediately, a buzzer shrilled, and the gate on my right side opened. Apparently, I "passed."

Feeling relieved but also very unwelcome, I rushed out into what seemed like a baggage area, started to breathe again, and ran to the restroom.

Bob's first impressions. I came to Russia for the first time several years later, but apparently not much had changed. The plane broke through the clouds in preparation for landing. It was winter, and Russia was still very much in economic peril after the fall of the Soviet Union.

I peered at the ground between the clouds. It was covered with snow; not much life. Occasionally a truck or two puttered along a narrow road. The trucks were old, often in ill repair, and when they had been repaired, it was in a makeshift, more-or-less functional manner. A metal panel might be welded over a hole in the cargo area. Eventually apartment buildings came into view – several big structures, maybe 20 floors, the kind you might see in a big city, but these were clustered together, more-or-less isolated from any town.

We landed, apparently on a runway isolated from buildings but bordered by a barbed-wire fence. It looked bleak. Eventually we taxied up to a building; another plane was nearby. When we stepped out of the door of the plane onto the stairs to the ground, Russian winter said "hello." It was cold, windy, damp, with a few snowflakes blowing in our faces. Brrr!

I walked with the other passengers into the building. Not much heat in there; no wonder, one wall was made of canvas and the wind was blowing in underneath and around one side. I needed a bathroom. I asked someone standing in the hall for the "WC." That did not work very well, so I tried "toilette." That did it. The gentleman pointed ahead and to the left. Soon I could smell it. Who knows when it had been cleaned last. There was no toilet paper or paper towels to dry one's hands. It didn't matter much; I was glad to get out of there.

My initial impression was a bleak, depressing, and poor country in ill-repair. It was the remnants of the political, economic, and social collapse of the Soviet Union (see Chapter 3). Over the years, it would recover in most respects.

Around town. We often walked around town to shop, always together for safety, but we never needed it.

St. Petersburg, for example, is a very interesting city. Historically and culturally, it is the heart of Russia. There are literally dozens of attractions that one can visit – it is, in our opinion, a great tourist town, and an expensive one. Despite the fact that most citizens are quite poor, it is among the most expensive cities for visitors in the world. It also has grandeur, which is best seen at night when the government, public, and other buildings are well illuminated.

There are dozens of famous places and buildings to visit, but the major sites include Peter's Palace (Peterhof) and Catherine's Palace (Yekaterininsky Dvorets in Pushkin), both out of town a short distance; the Hermitage Museum; and several churches, especially the Church of the Resurrection (or Savior on Spilled Blood) and St. Isaacs.

In the summer, a *boat trip* on the Neva River at midnight in June when the sun barely sets at all (called "White Nights") provides a fantastic experience to ride beneath the large bridges spanning the river that are wonderfully illuminated. Then, cruise under smaller bridges spanning the city's many canals and wave to the vodka-drinking folks overhead celebrating White Nights.

Peter's Palace is located 18 miles west of town. Peter the Great was quite a man. He wanted a palace patterned after Versailles, and he got it. The grounds are extensive, and the jewel is behind the palace, in its "backyard," consisting of two parallel cascading waterfalls decorated with fountains and golden statues. The water converges to a canal that empties into the sea. Pictures of this often adorn travel brochures. The amazing thing is that Peter himself designed all the fountains' hydraulics, all without pumps and motors. They still operate using naturally available water and gravity. The grounds and fountains are spectacular, far outshining the palace itself, in our opinion.

Catherine's Palace's appeal is just the opposite. Catherine, wife of Peter the Great, ruled for only two years, and the palace's grandeur is due mainly to her daughter, who used it as a summer home. Located 15 miles south of town, the palace's grounds are extensive and very nice, but it is the palace itself that is world-class impressive. Photographs show what the palace looked like after World War II – only portions of the main structural features remained along with a great deal of rubble. After years of reconstruction, the palace is now nearly as it once was. Here the palace itself seems modeled after Versailles, including a giant ballroom and room after elegant room, each decorated differently and in different colors, including the famous Amber Room.

The Hermitage Museum is located in the city on the Neva River. As an art museum it ranks up there with France's Louvre and Italy's Uffizi Galleries. It has a gargantuan collection – three million pieces in total, including whole rooms of Rembrandts and Van Goughs, for example. But it is easy to concentrate on the art and miss the building itself, which was the winter palace of the Romanov Tsars. It's immense and elegant. And the incredibly large courtyard and entranceway (Palace Square) was not only where visitors arrived but it was also the scene of gatherings that were part of the Bolshevik Revolution.

Across the Neva River lies *Vasilyevsky Island and SS Peter and Paul Cathedral.* The Cathedral is magnificent, decorated in white marble and gold, and many of Russia's rulers are buried within it.

We had not thought much about churches in a country that for decades did not encourage traditional religion. But many people remained quietly religious through the Soviet era, and their churches were preserved, and some were restored. One was the site at which the Bolshevik Revolution was planned, and there are dozens of other churches in different styles throughout the city. Two stand out for their unique architecture.

The *Church of the Resurrection* (also called *Savior on Spilled Blood)* was built on the site where Tsar Alexander II was murdered in his carriage in 1881, and a small section of the street where the assassination took place is preserved inside. The outside of the church is brightly multicolored, and pictures of it frequently advertise the city in travel brochures. But inside is miraculous. It was heavily damaged by Soviet vandalism, and it took 27 years to painstakingly restore the mosaic floor, walls, and ceiling – all the inside surfaces are covered with tiny mosaic pieces. Christ, in mosaic, looks down at you from inside one of the onion-shaped spires. It is a wonderous sight.

Just as the Church of the Blood is delicate and artistic, *St. Isaacs Cathedral* is monstrous, imposing, strong, and heavy. It is large, and many surfaces inside are covered in malachite. The exterior is not very "church-like," but the interior is quite traditional in form although unusual in decoration. There is a dome, and outside statues around the perimeter look down at people on the ground. Energetic folks can climb up to the top, from which one gets a good view of the city.

As indicated, many of these sites required an immense amount of restoration after the war, costing large amounts of money and taking decades to complete in some cases. We asked a few Russians what they thought of spending such large sums on these relics when most Russians were rather poor. Generally, the answer was: "These sites are our history, culture, and heritage. We need them to define our past. They also bring tourists and boost the economy."

Part of that history and culture are the artists that Russia has produced through the years. In St. Petersburg, there is an *artists' cemetery (Tikvin Cemetery)* on the edge of the Alexander Nevsky Monastery. It is much less visited than the sites described above but equally impressive in a different way. It's a very modestly sized area, and early on it was not well kept. It is divided into separate sections for composers, performers (often ballet stars), writers, and others. The spaces for the burial vaults are quite narrow, and they are placed very close to one another. We did not recognize most of the people buried here because of our own ignorance of Russian art, but we did recognize a few of the writers (Dostoyevsky) and most of the composers. Here were Tchaikovsky, Rimsky-Korsakov, Mussorgsky, Borodin, Glinka, Stravinsky, and more, all right next to each other. There must have been 15–20, all recognizable and famous composers. What a legacy of music, from one culture and mostly from one era! We had no doubt that those honored in the other sections were just as famous; we were just uneducated. The cemetery showed Russia's value and respect for the arts. When we rejoined our Russian colleagues after visiting the cemetery, we remarked on what a wonderful tribute it represented to these people who had contributed so much to Russia and the world. One of our colleagues commented soberly, "Yes, we know how to care for and honor the dead; it is the living we have trouble with."

Winter. Winter could be a bit exciting. St. Petersburg was quite pretty when snow covered everything, but there was no place to put the snow. Plows would pile it up, and then it had to be loaded onto dump trucks and hauled away.

More dangerously, at least for us, was walking in the winter. St. Petersburg is built on a former swamp, and because of the unstable land, buildings are restricted to a few floors. They all have flat roofs. In the city, the buildings abut the sidewalk, and the sidewalk extends from the buildings to the street. The buildings often have an inner courtyard, which is accessed by a narrow driveway for cars that runs from the street, across the sidewalk, through an opening in the building, to the central courtyard. At the driveway, the sidewalk dips to the level of the street and then back up to the level of the sidewalk.

Now about those flat roofs. When the snow melts, two things happen. First, giant icicles form on the edge of the roof directly over the sidewalk. Occasionally, they fall – squarely on the sidewalk a few feet from the edge of the building. Sometimes, parts of the sidewalk are cordoned off due to falling icicles, but of course a few icicles fall before they decide there is an icicle danger. We were told some people are injured or killed each year from the falling ice.

Second, the melted water on the roof is transported to the ground through large pipes roughly a foot in diameter that run down the front of the building. It does not get piped directly under the sidewalk into the gutter or the sewer. Instead, the drainpipes open directly onto the sidewalk. Some of the water gets to the gutter, but much of it forms a sheet of ice on the sidewalk. Somehow, through practice, culture, and maybe genes, the locals are totally undisturbed by this. They dart across the ice as if it were not there. Not the Americans, at least not us. The sidewalk could be very slippery, and we had to hang onto each other to get across these ice fields. Further, when we encountered a driveway that was iced over, we had a downhill ice skate followed by an uphill ice climb. And if the sidewalk was cordoned off because of icicles, we gingerly went around it in the road, which was unsafe for other reasons. We never fell, thank goodness, but we must have been entertaining to the locals as we slip-skated down the sidewalk. Maybe this is why Russia is good at ice skating and hockey.

Again, over the years, things improved – sort of. They progressively replaced those big drainpipes. That is, they replaced the old pipes with new pipes, but they still drained onto the sidewalk – too expensive to run them under the sidewalk, we suspect. Some things are hard to change.

The local economy. It was difficult to get a grasp of the economy, especially how people managed to live. In the 1990s, Russia was still in desperate economic circumstances, as described in Chapter 3. Unemployment was common, and salaries were meager for those who were employed. There was not much of a meritocracy – professionals, doctors, and professors, for example – seemed not to be paid a great deal more than basic service personnel. Salaries of $200–$400 per month were common. Yet, the cost of food seemed more-or-less equal in dollars to US prices. While apartments may have been inherited from the Soviet state and heat and electricity provided, we nevertheless asked our colleagues how people manage. They said, "As best they can. Most people eat cabbage and potatoes; meat and fish are too expensive."

People also seemed to hold two or three "full-time" jobs. We never figured out how they managed that, and maybe they didn't. There is an old saying, "We pretend to work, and the government pretends to pay us." Although they laughed when they said this, there might have been some truth to it. At the very least, they took money from whatever employment they could get – a kind of Russian "gig" economy.

People certainly lived frugally and modestly. For example, a pediatrician, her husband (an engineer), and an adolescent daughter might live in a communal apartment. In Soviet times, the large mansions of the rich were taken over by the state and transformed into communal housing for numerous families. The apartment was called "the railroad car." It was an apt description. They lived in one room, very narrow (perhaps 8 feet at most) and rather long (maybe 25+ feet). That was it for living, eating, and sleeping for the three of them. The apartment building had communal bathrooms that were very basic and in ill repair and a common kitchen area with perhaps five stoves/ovens and a sink and counter for each all in one room. This was shocking and enlightening. If this was where a pediatrician and engineer lived, how did the "ordinary" folks live? They often lived quite a distance from town. No wonder some of them wanted to work 24-hour shifts and then be off three days so they would not have to pay the subway fare every day and could work another job on their days off.

Because Russia has serious winters and a Soviet system that had provided the basics for most people, Russian poverty is not as obvious as it is in some other countries. Many people live in those isolated, very large, multistoried apartment complexes; few live in stand-alone houses. Although some of these buildings were old and in ill-repair, they represented places for people to live.

Over time, the economy improved but unevenly. Wages sometimes increased, but inflation could run 10%–14% per year. Eventually, bakeries had bread, stores had merchandise, vehicles and buildings were repaired, and things returned to more-or-less "normal."

A poignant history. World War II was especially hard on Russia, in particular the Siege of Leningrad, which is what St. Petersburg was called at the time. Some have speculated that Russia's plan to defeat the Nazis was to slowly let them into Russia and freeze and starve them out, but this strategy was horrifically hard on Russia. The Nazis eventually surrounded Leningrad, laid siege to it, and isolated it for two years, killing and starving hundreds of thousands of Russian citizens and soldiers.

Outside St. Petersburg is a cemetery commemorating those who died (Piskariovskoye Cemetery) during the Siege. There were dozens of mass unmarked graves, literally raised flat-topped mounds perhaps ten by ten meters square, containing the remains of countless casualties. At the back of this area was a stone stage containing a statue of "Mother Russia" while mournful music played across the entire cemetery. Not a word was spoken by anyone visiting the cemetery or a small museum. It was a dignified, horrendous, and emotionally moving sight.

The Siege took a more personal tone for us when we gathered with some Russian friends over dinner at one of their apartments. One of the guests was an older woman who spoke English among other languages. She was a young girl, perhaps nine years old or so, during the Siege. She recalled running over the main floors of bombed-out buildings to get provisions dropped by Allied planes to try to keep people alive. Her mother warned her not to go near the stairs leading to the basements of those buildings, because deceased victims of the bombing could still be lying there in the rubble. Occasionally she had to jump over people laying on the ground, perhaps dead. The provisions often consisted of dried grain and cereal, and her mother emphatically insisted that she not eat any of it but bring it home. She warned that many people were so hungry that they just started to eat it right out of the bags; but then they got thirsty, and they died either from impure water or the water expanded the dried grain beyond the size of their stomachs.

Chris then described her mother's story, equally horrific, of being rounded up by the Nazis in a town in Poland in the middle of the night. The Nazis loaded her, her mother, and brother into a truck and burned their house down with her grandfather inside. She described the retched conditions of the Siberian concentration camp, the minimal food, the bugs, disease, and the death of her brother one night laying next to her on the floor that passed for their beds.

The Russian people have endured hardship nearly throughout their history, although the war was clearly one of the most difficult epochs. Yet, despite his treatment of the Russian people and his killing of untold numbers of them, contemporary Russians regard Stalin as their country's hero, because he won the war against the Nazis. The Russian people are indeed long-suffering.

Russian people. Some vignettes provide a few insights into other aspects of the Russian people.

- On one of our earlier trips, we were met by an associate of the organization that had arranged our visit. She took us to a café upon arrival, ordered coffee for all of us (yes, coffee, no vodka), then looked at us and said, "Well, why are you here?"

  Bob was a bit shocked. We were to meet with a variety of people, including service administrators, political figures, scholars at the university, etc. He thought this was all arranged in advance. Chris, always the diplomat, calmly described our purpose and the people, or types of people, we wanted to see.

"OK," the associate declared, and she immediately started making calls on her mobile phone, and within 30 minutes she had several appointments set up over the next several days. "Anything else you want to do?"

This would not be the only instance in which things seemed to be done on the spur of the moment, at least by some individuals. Bob speculated that advanced planning was not a common activity, at least not for some Russians.

- We had an American colleague who lived in St. Petersburg for over a year. One day she was out walking on the street shortly after the 9-11 terrorist attacks in the United States. She came across a Russian mother pushing a stroller with a small baby in it. Our colleague paused to greet the mother and view the baby. Although she spoke Russian, her accent signaled she was foreign, and the mother asked where she was from. She said the United States. "Ooohh," empathized the mother with considerable emotion. "I'm so sorry for Americans. We Russians know such suffering," and she gave our colleague a big Russian bear hug of sympathy and kindness.

- Despite their relative poverty, we found ordinary Russian people honest and giving. In a small shop, Bob made a purchase and gave the clerk several bills to pay for it. The clerk started for the cash register but then turned around and immediately returned one of the bills to Bob. Then, he got change for the remaining bill. "He could have kept that bill," Bob thought, "and I would not have known that I gave him way too much for the item." This was not an isolated incident, because some weeks later a clerk chased us after we left the store to also return some money to us. Even an article we read said that ordinary Russians think the basic premise of capitalism – that the store sells an item for more than it paid for it – is cheating and dishonest.

- On what was Thanksgiving weekend in the United States, two service workers bought our American colleague a turkey, roasted it with vegetables, and delivered it all in a pot to her at her apartment. They had carried it on the subway for an hour, then walked with it for 30 minutes from the subway to her apartment. Our colleague was overcome with appreciation. "These people cannot afford that turkey plus the subway fares," she observed. "Then, to carry it through the Russian winter to me was so caring. That's how generous and loving these people are."

- At the end of her stay, our American colleague threw a party for her Russian friends, their spouses, and us at a restaurant a few blocks from her apartment. We all celebrated, a bit too much actually. Shots of straight vodka can do you in very quickly. Chris and Bob and the husband of a young service worker had to essentially carry one of the partygoers to her apartment. On the way back to the party, the young man said, "I was in the military. We were taught to hate Americans and that they were evil people. But you are not that way at all. You are very caring. I like Americans." Chalk one up for diplomacy, one person at a time, and remember when working abroad that you and your staff are representing your country to your international colleagues.

- One day, an American couple visited and joined us and a few of our Russian friends for some informal socializing. At one point, the conversation quickly turned to the adopted children of the couple and those of Chris and Bob. There were joys and fears and incidents both comical and disastrous – all those things American parents talk about when they get together and talk about their children. Our Russian friends were totally silent through it all. Afterwards, they said, "that was amazing for us. Russians would never talk about such things with each other, let alone with strangers. It is not just the 'secrecy of adoption' in Russia; Russian parents just would not do this. You Americans are so open about your personal lives. Amazing."

Our colleagues. A major lesson we learned across all countries is that we needed local colleagues to advise, guide, supervise, and participate as much as possible in the local project. The success of our projects was proportional to the extent of engagement and skills that local colleagues brought to the project. Further, it helped to know the extent and nature of their training and backgrounds. A degree in psychology, for example, can reflect very different knowledge and training in different countries, and these differences may not be apparent at first but only after you are well into the project. It helps to explore these similarities and differences as early in the project as possible, because differences in background can contribute to conflicts over the course of your project.

Latin America

Our work in Latin America took place in several different countries. Of course, each country is somewhat different, but there were general lessons learned.

Arrival and first impressions. In one country, arriving at the hotel for the first time was a little disconcerting. The hotel sat within a walled compound. The gate was opened for our car to enter by two guards armed with rifles. Inside, rifle-carrying guards were stationed in the lobby. During the day, armed guards went with the maids as they cleaned the rooms. In town, nearly every store and gas station had a rifle-toting guard to greet you. This country had a civil war that ended a few years before, during which citizens did not know who was on which side and youth roamed in gangs fighting and looting. Further, some of the legislators owned the security businesses and insisted on round-the-clock security protection for every store. Everyone was very friendly and took it all for granted; no one seemed concerned. The guards and guns were part of the furniture, but it helps to know this history and circumstances.

Our benefactor. Nearly all our work in Latin America was to evaluate interventions that were designed and funded by the President of a new American foundation. The President of the foundation was a very energetic, persistent young woman on a mission to improve care for vulnerable children in low-resource countries. She had limited formal training in child development, education, or psychology, but she knew what she wanted to do. She knew children needed more engaged, sensitive, and responsive caregiver–child interactions.

In general, we were on the same page as the Foundation President, but the differences in background had to be managed. First and foremost, we were hired to conduct an evaluation of HER intervention; we were not hired to help design the intervention. Second, she was very much a top-down manager in that she preferred to meet with the President of the country, his wife, a major national minister, and the US ambassador and receive a blessing from the top. We were often included in these meetings to provide independent credibility to the enterprise.

In contrast, our approach tended to emphasize more of a bottom-up strategy in which we would first work with the Agency's Director and senior staff and then the service personnel to enlist their support for an intervention. The Foundation President had us present a workshop for several Agency Directors and senior staff whom she wanted to implement her intervention. For the first half of the workshop, we presented some basics of child development, attachment, and related topics. At the break, she came up to us and said, "Why did you go through all this? Just tell them what we want them to do."

Chris responded, "Wait. There's a reason for presenting this material. We'll get there."

In the second half of the workshop, Chris listed six characteristics of a good family, and the Directors agreed that the family was the ideal context for rearing children. Then, she listed for each characteristic what it was like for children being served by these agencies. The service environments were exactly opposite on every attribute. Then, she related those characteristics to the general developmental needs of children that she had presented in the first half of the workshop.

The Directors nearly rose up in unison, "We have to change our services! We need to be more family-like." It was a great example of getting the Directors to design the intervention we and our benefactor wanted – a minor professional triumph.

For the most part, both the top-down and bottom-up approaches are useful and differ primarily in emphasis. It was not a problem. More difficult was that we had quite a bit of experience designing and implementing an intervention with effectiveness data to support it. But we were hired to conduct an evaluation of her program, not to help her design that inter-vention. We had to constantly remind ourselves of our role.

Local colleagues. As evaluators, we needed local colleagues to identify people who could be our assessors and to supervise and support them, because we would not be present the vast majority of the time. Initially, this was very successful. We were referred to a professor at a local university who was a psychologist, experienced in conducting empirical research, and a priest (see Chapter 9). But someone with this background is not always available in some countries.

Indeed, when the projects moved to another country, we had to find a local expert in that country. We did find one at a local university, but after some weeks she resigned from this role. We were not sure why – perhaps she did not like the duties, her university objected to this arrangement, who knows.

The unexpected. In a separate project and country unrelated to the above, we were invited by a former graduate student, now a professor at a local university, to present our work in a colloquium that was open to the public. We were also interested in exploring creating a project with our colleague in that country.

During the question-and-answer period following our presentation of our Eastern European intervention project, a woman stood up in the back of the audience and shouted, "How dare you come here and present this work to us! We don't want to improve institutions; we want to get rid of them. Every child deserves to be in a family not an institution of any kind."

That shot dead silence throughout in the auditorium for a couple of seconds. Then, Chris responded. "We agree. Ideally children should be reared in a warm, loving, stable family. There have been countries that have created family-centered services for vulnerable children, but it takes time and resources. Still some children may be hard to place and will remain in institutions. So we favor your efforts to create a family-based system. We just feel that one should provide the best care possible for ALL the children, including those who must remain in institutions."

This answer, of course, did not placate the advocate, who came up afterward to continue her message. Our colleague later apologized for this confrontation, but we told her we had heard this argument before. She indicated that this person headed a rather powerful group of advocates for family-centered care, among other causes.

Our colleague told us that she had invited the city administers to come to our presentation, but most did not. She was disappointed, because not much happens in the city unless those administrators do it or sanction it. However, one did come, and she was sufficiently impressed that she convinced her supervisors to invite us to make a presentation to a regular meeting of all of the city administrators who pertain to child welfare. The meeting is tomorrow morning! Would we agree to do it?

"Sure," agreed Chris. "Let's have a few more details." They wanted a somewhat different type of presentation, so we planned it that night and presented it the next day. The audience, it turned out, was not only made up of city administrators but almost 200 leaders of government and private services for children. The group was very receptive, and several stopped by at the end to ask questions and thank us for our presentation. While nothing happened as a result of this presentation, one needs to be prepared for the unexpected and willing to take advantage of opportunities when they arise.

China

Our work in China was greatly facilitated by our own staff member who was born in Shanghai. He made contact with a faculty member at a Shanghai university who was interested in early care and education and children with disabilities and who had very good relationships with many of the national government administrators. Our staff member was head of our division of applied research and evaluation, and he was an adoptive parent of two Chinese girls, so he was professionally and personally relevant to our work in China. He also planned and escorted our trips to China, and because he spoke Chinese, he mediated between us and the Chinese culture, institutions, and professional authorities. He was an enormous help. In addition, we had a graduate student born in Taiwan who was also very helpful. It may be an advantage to look for graduate students from the country you will work in to assist on the project and travel with you.

Public health. On our first visit, our plane landed in Shanghai, and passengers were told to remain seated even after the plane stopped at the gate. Then, suddenly two Chinese individuals boarded the plane. They looked like Martians. They were dressed in white over-suits and had head gear that looked a bit like gas masks. One started down the aisle spraying something into the air above our heads. Then, he returned to the front of the plane, and both of them held in their hands a curious gun-like gadget that they pointed at each passenger as they walked down the aisle. Once through the entire plane, they returned to the front and departed down the jetway. They were disinfecting the plane, and the handheld "gun" checked each passenger for a fever. We were now free to commence to passport control. Some countries are concerned about public health, and this was years before the COVID-19 pandemic.

Shopping Chinese style. Chris had a friend in Shanghai who was highly placed in a business firm. She arranged for us to rent large apartments for our visits, and she lent a colleague to help guide us for an afternoon of sightseeing and shopping. We visited a large open market, and he did the negotiating for our purchases. He was incredibly persistent, forceful, and aggressive, even insulting and threatening the young girls who were the salespeople, bringing one to tears. We couldn't watch – let us know when you've agreed on a price, and we will pay it. It was a great contrast in style to our stereotypical impression of Chinese social relations. Our Chinese-born American colleague told us this is the way purchases are made in

China, but even he thought our guide was too aggressive, perhaps wanting to get the best price for the American guests of his boss.

The next day, we went shopping on our own on a street in Shanghai loaded with one jewelry shop after another. Chris, a native of New York City, was always a good bargainer but also a diplomat, and she was a little emboldened by her experience the previous day. She managed to get what she thought was a good price for her purchases, and there were no tears and everyone seemed happy at the end. A little local diplomacy one shop at a time.

Culinary adventures. One of the pleasures – and adventures – of working in another country is sampling the local cuisine. Chinese food is among the great cuisines of the world. But they do eat some unusual things occasionally that are someplace between strange and repugnant to Westerners. For most of our work in China, we were with our Chinese-born US colleague, himself quite a good cook, who guided our selections at most restaurants. Even then, some of our US colleagues found the food "a little unusual."

On occasion, so did we! On one trip to China, we were alone and hosted by two Chinese scholars who took us to a very "local" restaurant. It was enormous, and we were the only Westerners in the place. Our guests talked about the specialties of the restaurant, one of which was a green soup that they said Westerners usually did not like. The meal proceeded well, and everything was quite good.

Then, a bowl of soup arrived. It was green!

"I think this is IT," Chris whispered to Bob under the din of noise in the restaurant. Our hosts offered it to her first. She spooned a little into her bowl, and passed it to Bob, accompanied by a clandestine whisper, "It smells like vomit."

Bob agreed. It did! We ate it, politely, while our Chinese hosts tried hard to conceal their smiles. More beer! Poor Chris does not drink beer, but it was better than the green something or other.

On other occasions we were fed – or voluntarily sampled – a variety of local delicacies, some common, many of which were delicious if not unusual to us, while a few others ranged from edible to wretched in taste or texture. A partial list of such "experiences" included jelly fish, whole sparrow, whole prawns, watermelon juice, eel, crab roe, squid, beggar's chicken (whole chicken encased and cooked in clay and smashed open at the table), sea cucumber, sea slugs, shark fin soup, beef tendon, duck neck, 1,000-year-old eggs, fish cheek, and pigeon with head. Be prepared to sample some unusual foods and smile diplomatically. By the way, a little hot sauce can mask some less appealing flavors.

Toe nibbles and foot massages. We visited a children's hospital in a Chinese city and gave talks to a conference there. In appreciation, we were taken as guests of the hospital to a private "club" that specialized in water activities of various sorts, including a swimming pool, whirlpool, and various "spas," one of which consisted of a small pool containing numerous little fish. One put one's feet in the water and the fish would nibble on them. The fish were quite gentle, and their "nibbles" felt like ants crawling on one's feet. Many years later, we learned that this "fish pedicure" originated in the Middle East and is spreading in popularity around the globe. The fish, called Garra rufa, nibble away dead skin, but it's health and safety is debatable, and it is banned in some US states.

Our American Chinese colleague very much enjoyed foot massages, and whole salons were devoted to the enterprise. We trusted him to take us to a reputable establishment, and he did. The experience starts with soaking one's feet in blisteringly hot water, followed by a rather vigorous massage. Chris and a few of our other US staff members enjoyed this experience quite a bit; Bob not so much. He found the foot massage quite painful, and the masseuse accommodated but later told him that such sensitivity was a sign of a heart problem. A few years later Bob had a double cardiac bypass operation. Mmmmm!

Kazakhstan

The country and its people. Kazakhstan is the ninth largest country by territory and the largest land-locked country in the world, but it is somewhat isolated. It lies in Central/East Asia, west of China, south of Russia, and due north of India and the other smaller "stans."

Let us tell you, Kazakhstan is a long way away from Pittsburgh – essentially 30+ hours portal to portal. Planes do not fly every day to the capital city Astana (now called Nur-Sultan), so you need to schedule carefully and leave yourself a long layover in connecting cities, so you don't miss the connection and wait two days for the next flight (see Chapter 6). Also, flights arrive in Nur-Sultan at 4 am to 5 am in the morning and depart at roughly the same time. The local hosts arranged a meeting the afternoon of the day of our arrival on our first visit. We could barely keep our eyes open and be coherent. Although flying East seems more difficult to us than flying West and we are getting older, we requested two entire days of adjustment before any work on subsequent trips.

Kazakhstan is a former Soviet state, so it has a substantial Russian population as a consequence of the Soviet Union's attempt to acculturate its allied states. We learned later in our work that there was some tension between Russians and native Kazakhs, and there was a desire to have American expertise and consultants to help develop projects there. Indeed, the University of Pittsburgh Medical Center is a major consultant to the president's university medical school in Nur-Sultan.

The local people are a mix of religious/ethnicities, with native Kazakhs being a blend of Chinese and Caucasian roots and of the Muslim faith whereas the Russians are more Caucasian and likely Christian. There are no outward signs of the Muslim faith in the dress of people, but they do not drink alcohol (but did not mind us having wine), and there are mosques located about town. Most Kazakhs speak Russian and Kazakh.

Politically, after the fall of the Soviet Union in 1991, Kazakhstan became an independent country. Its president is elected every 5 years, and its first president had no term limits and only very recently stepped down unexpectedly. He established the capital in what was then Astana, not in the largest city of Almaty, and basically built a new city next to the old. The new city has wide streets, creative architecture, a magnificent mosque, and a modern university named after the president in which all classes are taught in English. Several businesses, especially construction companies, have flourished building the new city and country.

The land is quite barren around Nur-Sultan, supporting the raising of horses but little else. Horses have a prominent role in Kazakh history and culture; they are revered, and horse meat is a traditional dish. We tried it; it is firm meat often served packed with fat inside a large sausage. It does not taste much different than full-flavored beef. Fermented horse milk is another delicacy with an ancient history going back to the Moguls, generally not favored by foreigners like Americans, including us. But when you are out to dinner with your funder, you try the local staples.

Winters in Nur-Sultan can be challenging. It can get quite cold, as in 20+ below zero Fahrenheit. It can also be quite windy, sending windchills to −40 degrees. Fortunately, we have fortuitously avoided the most severe weather. Summers are pleasant and not too hot.

Our colleagues. We had two types of colleagues. One was Katrina, the Director of the Pittsburgh adoption agency who had stimulated several of our international projects. She functioned as our "agent" in Kazakhstan. Not only did she arrange for our first trip to explore possibilities in Kazakhstan, but later she invited several of the foundation women to Pittsburgh to meet with us, and subsequently she acted as an informal intermediary between us and the foundation administrators.

Our second colleagues were wonderful, devoted women of stature in their country. As a group, they represented many affluent and influential families in Kazakhstan, and they funded several organizations and services for vulnerable children and their parents. Although their foundation was newly organized and still formulating its program's direction, they were intent on helping needy children and parents in Kazakhstan achieve better lives.

The foundation was not affiliated with the government, but the women knew all the relevant government officials. There were advantages and disadvantages to this arrangement. For example, they had ready access to financial resources and many fewer strings attached to how it could be spent. On the other hand, the government controlled many of the services and access to them.

Our Kazakh colleagues showed us a variety of services for vulnerable children. We visited a home for older children, a day clinic for children with severe disabilities (e.g., cerebral palsy), and a Mother's Home, which was similar to our old halfway houses for pregnant and new mothers to live and receive support to help them keep their babies rather than relinquish them to institutions. The foundation or some of the women personally financially supported these services.

We were assigned a Kazakh colleague who worked for the foundation and who mediated everything for us during our projects there. She and especially her daughter spoke English (as did some of her colleagues at the foundation), and she arranged our visits, monitored the project, and obtained data for us. We met occasionally with the foundation executives, including the principal donor (the President of the country's main construction company), his wife (the President of the foundation), the foundation's Chief Operational Officer, and several other women active in the foundation. They were all very committed and supportive.

More politics. Early on we met with some government officials with our foundation representatives and described our work and the improvements it produced in children's development. One official said their institutions, for example, were doing very well; they and their children did not need any improvement. On a recent visit to an institution the official reported that "the children came running up to me, a stranger to them, and gave me repeated hugs. They were so glad to see me they hung onto my skirt as I toured their room. They were happy and doing fine."

We bit our tongues and refrained from telling her that those were signs of "indiscriminate friendliness," which is often exhibited by extremely neglected children and that no parent-reared child would do those things with a complete stranger. Instead, we assured the minister that children in Kazakhstan's institutions were not doing well in physical growth and mental development. The response was, "How do you know that? You made that up." Fortunately, we had the published article written by other US scholars that documented these facts obtained from ten institutions across Kazakhstan, including in Nur-Sultan. We gave it to the government administrators, but the article was in English, and we doubt they ever read it. But our foundation colleagues appreciated that we provided independent confirmation of what they had been telling the officials and brought American science to document it.

The robbery. One of our major activities in Nur-Sultan was to create and present with additional colleagues from the United States and the Russian Federation a seven-day workshop for Kazakh professionals on improving care for vulnerable children. Approximately 25 diverse specialists attended, and the sessions were held on the sixth floor of an eight-story building owned by one of Kazakhstan's major construction companies, the President of which and his wife were the major contributors to the foundation sponsoring our work. The building was quite modern and well-appointed, with a nice lobby and fairly tight security arrangements prior to proceeding to the elevators and beyond.

After one morning session, the group broke for lunch, which was served in the second-floor cafeteria. Participants left their notes, handouts, coats, and personal items in the workshop room, and moved down to the cafeteria. After lunch, people filed back into the room a few at a time. Then, one of the first women to return suddenly shouted, "They're gone! They are gone! They were in my purse, but now they are gone."

A few women gathered around her, as she searched through her purse again. "Someone has been in my purse!"

"What are you missing?"

"Diamond earrings and a neckless that my husband gave me this morning. Today is our wedding anniversary."

"Oh dear, that's terrible." Everyone hurried to their personal items and searched their purses. "Some money is gone from mine," another anguished participant yelled.

"Mine, too!" joined another.

The Chief Operational Officer of the Foundation, who was attending the workshop, returned from lunch and immediately took charge. "Search your belongings" she ordered the participants as she called building security.

Then, Chris returned, and Bob immediately urged her to check her purse. "I did not leave this interior pocket unzipped – I never do," reported Chris. "Here's my wallet, and…," as she went through it, "Mmmm, I think some money is missing. I certainly had much more than this in here."

The Foundation's Operational Manager rushed over. "How much are you missing?"

"I'm not really sure, maybe $100 in US currency, maybe less….maybe more, I really don't know."

By this time, everyone had returned from lunch and checked their purses and belongings. Many said someone had been in their purse, and others reported money missing. The Operational Manager left to call building security, and shortly she returned to announce, "Security is on top of this. The police have been called. The foundation will replace all the money that has been stolen and the value of other items taken." After much commiserating among participants, the group decided to continue with the training.

Bob thought it was odd that no security official or police arrived immediately at the room to investigate. Was this an insider who did this; otherwise, how did he get past security? The Foundation's Chief Operational Officer came in and went out several times; clearly she was on this case. About two hours later she announced, "The police will be here in a few minutes. They will want to talk to several of you privately." Then, she came over to Chris. "They especially want to talk to you, Chris, and Bob, too. They are very concerned about Americans being one of the victims of this theft."

"Why is that?" asked Chris. "I only lost money. Our colleague lost something perhaps more valuable and a memento of her wedding anniversary. That's awful."

"No, no. It's not the value of items taken. It's that Americans were victims. About a month ago a wealthy American businessman was robbed of a very large amount of money. It made all the newspapers, the President apologized, and it was an embarrassment to the country. We don't want anyone to get robbed, but we also need and want American business to feel welcome and comfortable here. Having another American robbed so shortly after that episode looks very bad, much worse than the amount of money involved."

Two police officers arrived a few minutes later, and they indeed wanted to talk with Chris privately off to the side of the training session, which continued. They were very polite, calm, and spoke English rather well. They wanted a great deal of information, not just about what was missing but about Chris personally – basic biographical information, addresses, email addresses, purpose in Nur-Sultan, where she was staying in town, how long, etc. While the interview was conducted in a very respectful and matter-of-fact tone, the extent of the questioning was almost more characteristic of what we would have thought would be done for a potential robbery suspect than a victim.

At the end of the day, the President of the construction company came down to apologize for the incident and assure everyone that the situation would be taken care of and losses would be compensated.

After he left, the Foundation Manager announced that the robber had been apprehended. He was not an employee, but he actually had committed a similar burglary in the same building a few weeks before. He got by first-floor security by mixing into a group when officials were busy, but his image was caught on the security videotape both coming into and going out of the building. He was recognized as the previous robber and immediately arrested at his apartment. The next day, the Foundation compensated all loses, but we did not hear whether the earrings and neckless were recovered.

Stuff happens, even in what would appear to be a very safe environment.

5 Obtaining Wise Advice

General Lessons Learned

Contact any offices within your home institution that support international projects. Although such an office did not exist for us when we started in the mid-1990s, facilities that support international work are now commonly available in universities, for example. They can provide a variety of institutional regulations you may have to follow and offer some valuable services. For example, they may be concerned about electronic theft and hacking while you are abroad, which could do serious damage to your institution's technology systems. They may loan you computers and phones for your trips.

If your project is considered research, then contact your Institutional Review Board very early in the planning stage. Ethical reviews of international projects are much more systematic than they were some years ago, and other countries may have their own regulations. This is especially true for the European Union, for example, which has very restrictive regulations on collecting data and having such information leave any EU country. If you intend to work in an EU country, you would be best served by having a colleague in that country who is experienced with these regulations be a major collaborator.

It's advisable for US professionals embarking on an international collaboration to consult with the State Department (see also Chapter 6) to obtain advisories on the intended country and to register with the State Department for each trip (www.travel.state.gov). The "advisories" detail all the visa, political, economic, social, health, and other issues going on in the country that might bear on your safety and your ability to function there (see below). You should also visit the Centers for Disease Control and Prevention website (www.cdc.gov) for health information.

These advisories will give you advice on many local customs, policies, and laws that are different than in the United States. For example, they may advise that only pristine US dollar bills will be accepted, you must register with the police for stays over three days, certain medical clearances may be required for stays over three months, and so on. US citizens are often unaware that the United States requires a variety of specific conditions for visitors as well.

It is also a good idea to inform your country's embassy or the consulate nearest to where you will be working about your project, funder, collaborators, and implications of the project for your country and for the country in which you are working. This is especially true if your project is funded by an agency of your government or a foreign government. Not only does the embassy or consulate need to know what is going on in the country, especially activities sponsored by its government, but they can be helpful to you in conforming to local laws and procedures and navigating local customs and potential issues.

Illustrations

A Visit to the US Consulate

A project we conducted in Russia was funded by the US government, and it began in 2000. As noted in Chapter 3, this was a time of substantial political and social-economic turmoil in Russia. While we had read State Department advisories before the start of the project, we felt it important to visit with the US Consulate shortly after it was underway. We requested a meeting with an official of the consulate and sent a brief description of our project and its funding plus short biographical statements about us and our Russian colleagues. We were pleased to be able to meet with a fairly high-ranking Consulate Officer.

After navigating elaborate security precautions overseen by a US Marine in full formal dress and introducing ourselves to the Consulate Officer, Bob opened with, "We are very pleased to meet you and thank you for taking the time to have this conversation."

"You have a very interesting and important project," the Officer responded. "Not only is it funded by the US government, but it is the kind of activity that we like to have here. We often spend our time putting out fires and smoothing things over with the Russian government, but this is something that both Russian and US policymakers can view with pride."

"Thank you, we hope so."

"In fact," the Officer continued, "we often have members of Congress who come to visit, especially in the summer, and this would be an ideal project for them to see. Would that be possible?"

"Well," answered Chris, "it's wonderful that you think the project merits that kind of attention. But the project just started, and the intervention is just now being implemented. So, there is nothing to see yet."

"Also," added Bob, "one of the things we wanted to talk to you about was whether we or our Russian colleagues were in any danger operating this project here? We have been told that someone was murdered on the street in front of the agency in which we are working by a roof-top sniper a few months ago."

"I don't think you are in danger," assured the Officer. "There are quite a few murders here and in Russia in general, but you do not have anything to sell, you will not threaten someone's business, you are not paying off anyone – I HOPE – and you don't have a political axe to grind. Those are the people who become targets."

"We understand," acknowledged Bob, "but the project is funded by the US government, and someone may make the assumption that there is money around – for example, that we might bring large sums of money with us and that our Russian colleagues and their agency may have some."

"That's a good point," agreed the Officer. "Don't bring much money with you into the country."

"We have to bring some with us to pay for our lodging, meals, and some other expenses, but most of it is spent the first day we are here, so we are not walking around with loads of cash. The bulk of the money to operate the project is not laying around the agency either. It is handled by the Civilian Research and Development Fund, which was created jointly by the US and Russian governments to handle research funding in Russia. They pay people, buy equipment, and so forth – neither we nor our Russian colleagues touch the project money. But importantly, other people don't know that. They may assume there is money around."

"That's great that the finances are handled in this way. By the way, it is not uncommon here for people to demand payments – extortion or bribes if you will. Please do not make any such payments. It is against US law for you to do so."

"Thanks for that advice, and we were aware of this possibility. We have agreed with our Russian colleagues that we do not pay people unless they perform a needed service for the project," assures Bob.

"What we worry about is the safety of our Russian colleagues," continues Bob. "As you probably know, some Russian scientists who have obtained grants from foreign sources have been raided by the secret police and their data and computers confiscated. The police were tipped off by jealous Russian scientific colleagues. Something like this could happen, or someone could try to rob our colleagues, even in their service agency, believing that they have some of the research money on hand."

"That kind of thing has certainly happened," admitted the Officer, "and it is sensitive of you to be concerned about that. But we at the Consulate cannot do anything about that. We can protect and advise you, but not your Russian colleagues."

"OK," Chris acknowledged, "but this possible situation is a reason why we would prefer that you not bring US Congress people or other government officials to visit the project, at least not until it is completed. We suspect that such a visit would be covered by the local media, and that might inform and stimulate someone to threaten the project or our colleagues."

"Mmmm, I suppose you are right to be cautious about that possibility.

It's too bad, because this is the perfect thing for US legislators to see and know about – it is a positive, humanitarian use of taxpayer dollars."

"We understand," said Chris. "Most Americans would probably view it that way, but some Russians might see it quite differently. They might say, 'We can take care of our own children; we don't need the Americans to tell us we are not doing it well enough and that the services for children and families need to be improved.'"

"Fair enough," lamented the Official, "I agree we will not bring government officials to visit your project. Too bad."

"Another situation we are concerned about," continued Bob, "is having the police or some official claim that we need to pay a fee or something to them. We hear that this happens occasionally."

"Well, yes," acknowledged the Officer, "but not that often. You do not owe any fees, except the registration fee on each trip."

"Yes, we know about that fee," added Chris. "The agency takes care of it the first day of each visit."

"Are you driving around town?" the Official asked. "A traffic stop is often when such extortion takes place."

"We are driven by the agency's driver," offered Chris, "but we personally do not drive a car here."

"Well, if a police officer or other official asks you to pay a fee," advised the Officer, "tell him you do not have the money with you and that the director of the agency keeps it for you. You need him to come with you to the agency to get the money. Chances are that the officer will drop the matter, because he knows that the agency director will know right away that his request is bogus and is simply an attempt at extortion."

"OK," agreed Bob, "that is very clever. Thanks for your advice and support."

"I am very pleased you contacted the Consulate," the Officer concluded. "You have a great project, and we are pleased to know about it. Here is the latest advisory statement on Russia. It provides a great deal of information on other issues you should consider as you conduct your project. Best wishes."

These on-line advisories (www.travel.state.gov) are intended to warn tourists, business people, and other travelers about potential dangers and how to avoid them. Given the state of Russia in 2000, the advisory at the time was pretty negative. It gives you some idea of the kinds of topics covered in these documents, and we add how we dealt with some of them. Note that these items are from the early 2000s, and they will be quite different now. They vary substantially depending upon the specific country, and they can change frequently, so these are only illustrative examples:

- *Travelers cannot access money easily, if at all, via credit cards or wire transfers.* We knew this, and we brought cash and had it converted into rubles with the help of our Russian colleagues immediately upon entering the country. Credit cards are now commonly accepted, and ATMs are available.

- *Banks want pristine US bills with no folds, markings of any kind, or pin holes.* Indeed, whenever we converted US dollars, the bank would reject some bills that were not perfect, run each bill under a scanning light, and hold it up to the light to detect pin holes. Rejecting worn or imperfect bills is against currency laws, but it is very frequently done. Trying to get a US bank to give us perfect bills to bring to Russia was often a chore, and we sometimes had to give the bank a day or two notice to get the cash ready for us. It's incredible how many US bills are marked, especially on the edges. It seemed to us that perfect bills would appear more likely to be counterfeit to the Russians, but the Russians claimed it was the American banks that wanted the pristine bills. We do not know the truth of this practice.

- *One needs a visa to both enter and to leave Russia, and one needs to be invited by a Russian individual or organization.* Our Russian colleagues arranged this, often using a hotel or other individual or organization that would do this for a fee even though they had nothing to do with us or our visit. We had to be careful that we had the name and address of these inviters, because it was requested on the entry forms when we arrived in the country.

- *If you will visit more than three days in Russia, you must register with the police and pay a registration fee of $30.* We paid the agency, which registered us with the police.

- *Visas are usually limited to a stay of no longer than three months, unless specially requested.* Our American colleague often stayed longer than three months. On at least one occasion her stay was shortened on the visa without notice or explanation. CHECK YOUR VISA DATES AND OTHER INFORMATION. At that time, stays longer than three months required an HIV test and certificate in both English and Russian. Sometimes we had to get tested either before or after arriving in Russia, even for shorter visits.

- *Have photocopies of your passport and visa with you and with a travel companion at all times.* This is in case your status in the country is challenged or your documents are lost or stolen.

- *Over the past several years, unexplained acts of violence, including bombings, especially in the Chechnya region, have occurred at Russian government buildings, hotels, tourist sites, public transportation, and residential complexes. Be aware of your surroundings at all times.*

- *Crime against foreigners is a problem, especially in major cities. Harassment of and attacks on foreigners of Asian and African descent have occurred.* Racial/ethnic diversity may be more limited in some countries. During our project we brought an Asian American to Russia to help with the database construction, and we considered using an African American as a consultant but decided against that, in part because of this warning.

- *Medical facilities and care are usually far below Western standards. Get a list of Western clinics and English-speaking doctors.* We happened to locate an

American Medical Clinic a few blocks from the agency in which we worked. We visited it to clarify procedures, such as would someone who spoke English answer the telephone, would they pick us up at our residence in an ambulance if necessary, how would we pay for any service, and did they take insurance claims? Be sure to check your health insurance for coverage in a foreign country before departure (e.g., US Medicare does not cover you) and how to pay for it. We placed the clinic's calling card near the telephone in every place we stayed. Fortunately, we did not need it.

- *Travel conveyances in Russia may not be up to Western standards of safety. Not all airlines are considered safe* (https://www.airlineratings.com>safety-rating-tool), *and criminals have robbed and attacked foreigners on trains, especially between Moscow and St. Petersburg.* On one of her early trips to Russia, Chris did take the train from St. Petersburg to Moscow. She was approached by a Russian demanding documents and money in Russian. Fortunately, a young woman who spoke English interceded and got him out of their compartment. Later, Chris and Bob traveled to Moscow, and we opted for air travel because of this warning. The safety records of some local airlines in some countries can be less than Westerners usually expect, so we consulted our advisor at the Civilian Research and Development Foundation for advice on which airlines were safe and reliable.

- *Be mindful of Russian customs regulations prohibiting certain articles from being brought into and taken out of Russia. Penalties for having drugs, for example, can be severe. Have all medications in their original pharmacy containers.*

After reading one of these advisories, it is easy to feel that traveling to this country is fairly risky. The purpose of such advisories is to warn travelers of potential dangers; they are not travelogues promoting the country. So, they can read quite negatively, depending on the country. But can you imagine what such an advisory for the United States, for example, might look like: "Occasional mass shootings in schools, movie theaters, at concerts, and in places of worship; occasionally people of African and Asian descent may be targeted for abuse or violence; keep a close eye on the weather in certain parts of the country that are subject to tornadoes, hurricanes, floods, and earthquakes that can kill numerous people."

Other Countries

Generally, we visited other countries under the auspices of our funders, and they were primarily responsible for us during our stay. Nevertheless, we consulted the State Department and Centers for Disease Control and Prevention websites and registered with the State Department for each trip. In China, we had our own Chinese American staff member as a guide, who made most of our arrangements. We trusted his judgment completely.

Although armed security guards were often present nearly everywhere in one country, we rarely had any problem with crime (but see Chapter 4), but one of us did need some emergency medical care in one country. Fortunately, our local contact's father was a physician and could refer us to medical facilities, as they were available. Medical care in some countries can vary in quality and how it is delivered. In our one experience at a public clinic in Latin America, which was very basic in appearance and had a crowd of local citizens filling the waiting area, payment in cash (US dollars) was desired and promise of that produced immediate attention of physicians over all the others in the waiting room. But no receipts for us to give to our insurance company would be provided. A few days later, we visited a private clinic, which was totally different and equivalent to Western care and facilities.

"Stuff happens," so be prepared.

6 Tips on Traveling Abroad

General Lessons Learned

International travel is much more complicated than a domestic trip, especially if you are going for professional reasons and you are employed at a university or other large organization. For example, some of the relevant practices and policies are different for different employers and different airlines, and some keep changing. This is especially true in the aftermath of COVID-19. It is wise to become familiar with at least some of these practices and with the travel landscape of the internet. Below are numerous tips stemming from our experience prior to COVID-19, but be mindful that policies, facilities, fares, and regulations keep changing. Check everything, frequently.

Months Before You Go (If Possible)

- **Check your institution's regulations** on which airlines you may use (if using a federal grant to pay for tickets, Americans may be required to use a US-based airline for the trans-oceanic leg) and which class of service you may use. Check with your institution on other requirements, such as registering your trip with your institution and obtaining information on securing your smart phone or perhaps getting loaned electronic equipment for your travels. Universities are particularly concerned about what data you are taking with you to avoid data theft and minimizing security breaches from your smart phone and other electronics. Be especially mindful of university regulations regarding taking students or staff with you.
- **Be mindful of how things are paid for in the countries you will visit.** Some countries (i.e., China) have gone completely digital (i.e., QR code) – no cash, no credit cards. Find out what can be used to pay for things in these countries.
- **Investigate your credit card benefits** (your employer may have a card for you to use). Call your credit card customer service number or go online, and tell them when you will be gone and where you are

going so they don't decide your first charge is bogus and cancel your card, leaving you stranded with no money. Having two cards (i.e., credit and debit or two different credit cards) is prudent.

- **Find out what charges will be levied for obtaining a cash advance at a foreign ATM.** A credit card may not charge you a fee for purchases but will charge a fee and/or interest on the cash advance money until your monthly bill is paid. Debit cards may not charge you interest because they take the money directly out of your account, but they may have a limit on the size of a cash advance (and so may specific ATMs).
- **Some credit cards offer trip cancellation insurance** up to a specific amount, and your institution may or may not permit you to charge a trip cancellation insurance premium on your institutional account. If you feel you need to buy trip cancellation insurance, investigate carefully exactly what it will and will not cover, and which coverage you are already provided by other sources and which you really need. Typical travel insurance policies only cover documented illnesses as a reason for cancellation. They don't cover non-medical reasons for cancellation, including pandemics and fear of pandemic illness. "Cancel for any reason" insurance covers you for more reasons (but not necessarily "any" reason), it may not pay the entire amount and is very expensive – but so is having to cancel a $5,000 to $10,000 trip. Investigate carefully.
- **Make sure your credit card pin number is only four digits long** (some foreign ATMs limit the pin to four digits). Generally, ATMs will be your best source of local currency for daily living. Moreover, you do not have to change money, and the exchange rate you will be given on your card is very favorable (although there will be a fixed fee for each transaction). If you will need to take much larger sums of money, especially cash, ask your employer how they recommend you handle such transactions. Banks will change dollars to the local currency for a fee, and the exchange rate may not be favorable. Also, some countries insist on extremely pristine dollar bills (i.e., no marks whatsoever; see Chapter 5). Obtain sleeves or holders for credit cards and passport that have RFID protection against electronic stealing of your information.
- **Americans should consult the US State Department website (www.travel.state.gov) and enroll in its Smart Traveler Enrollment Program (STEP),** which will provide information on entrance requirements for the countries you are visiting plus internet warnings of political unrest, weather threats, etc. It also will provide the email, phone number, and street address of the nearest embassy or consulate in case of a problem during your trip (see Chapter 5).
- **Be especially mindful of any visa and other documents required by the countries you will visit.** The US State Department can provide this information, but also consult the websites of the countries

you will visit. Be attentive to non-visa documents you may need (e.g., birth certificates, especially for minor children; immunization records; documentation of negative COVID-19 tests and vaccinations; prescription drugs; baggage allowances in different countries). Even the European Union requires that you obtain a "travel authorization." Recheck these items closer to the time you will depart because they keep changing. Carry photocopies of all documents with you, and give copies to a travel companion (including credit card numbers and phone numbers in case they are lost). Take along extra copies of your passport/visa pictures, and identify a person in your home country who has all your information and can help in case of a serious problem. If you live near a country's embassy, you can probably get their visa on your own, but it can take time. Otherwise, use a commercial visa company. Be aware that some countries require a great deal of information on a visa application, especially about all your previous foreign travels. Also, if you are going to make this trip several times in a year or so, investigate multiple entry visas, which some countries (but not all) offer at less cost and inconvenience than getting separate single-entry visas. Do not delay; visas on short notice can be quite expensive.

- **Make sure your passport is valid for at least six months** beyond the scheduled date of your return (it could be longer in some countries; check with the State Department).

- **Americans should consult the Centers for Disease Control and Prevention (CDC) website** (www.cdc.gov/travel/destination) for the countries they will visit to obtain vaccination requirements as well as health information and recommended local English-speaking clinics and hospitals. Start investigating vaccinations early, because some require two doses with weeks in between. Have all this information, a list of medications you take, and important health records handy during your trip. If something happens, you may not have time to get this information. Consult your physician regarding medications you should take along in case of a bout of traveler's disorders, and obey the suggestions for what not to eat or drink (especially the water). The CDC will also provide safety conditions and warnings. These warnings can sound quite ominous, but it's helpful to be prepared and aware.

- **Be as informed as possible about regulations, procedures, and restrictions associated with COVID-19.** For example, go to *covidcontrols.co* for information on policies in each country and airport you will visit and *airsiders.com* for information on connections, safety, and restrictions at each airport. These policies may involve required COVID-19 testing and vaccinations, quarantines, difficulties connecting between certain airlines, etc. These regulations keep changing, so check early in your planning and also near your departure date.

- **Driving a car in some countries can be quite challenging.** It is probably best for you to use local professional ground transportation in the country in which you will work. Most (but not all) Western European countries have relatively low accident and fatality rates, but they can have a few road features that may be unique to your experience (e.g., roundabouts, driving on the left). However, many other countries have much worse safety records even for local drivers but especially for out-of-country drivers. If you expect to use a good deal of ground transportation, and especially if you will drive a car in the project's country, consider becoming a member of the Association for Safe International Road Travel (ASIRT; www.asirt.org) and obtaining their Road Safety Reviews.

- **Consult your health insurance** regarding what it will and will not cover while you are away and what procedures you need to follow to qualify for reimbursement. Medicare will not cover you outside the United States. If you are not well-covered, consider travel insurance, but it's quite expensive (about 8%–10% of the cost of the trip – see above). Note that in some countries, doctors and clinics are eager to serve you in preference to locals if you pay cash, and they may not issue a receipt (the fee may be sufficiently modest to be worth foregoing a reimbursement). American clinics and major international hospitals are likely to be more business conventional and medically competent.

- **Visit the internet, and study the country's local customs,** such as how to greet people (Do you shake hands or not and with whom? How do you greet political or "royal" dignitaries?), should you make eye contact or not, do you remove your shoes when entering houses/ apartments, what are the tipping practices (tip, no tip, for which services, how much), what are gift practices and, if common, come prepared with a variety of gifts for different types of people, and other customs.

- **If you are going to do this trip frequently, Americans should consider getting a Global Entry (GOES) card,** which also provides you with TSA pre-check privileges. The GOES entry into the United States is often much faster, does not require you to fill out the US customs form (you may do it at a computer upon entry), and spares you US customs inspections and questions. It costs $100 for five years, but you should apply several months in advance (it can take several weeks to get the required in-person interview at an airport that has a customs office).

Airline Tickets

- **Using the internet** to buy domestic airline tickets is reasonable and convenient. You can buy international tickets online, and if you are going to conventional locations (e.g., Western Europe), that also may be convenient. Recognize that fares are changing constantly, sometimes right before your eyes, and they can be much cheaper for flights on certain days of the week. If you use an online travel site (e.g., Kayak, Travelocity, etc.), also check the website of the airline you intend to use, because they sometimes have cheaper fares and sometimes there are perks that you get if booking through the airline that you do not get when booking through an independent site. Some airline websites provide this information. But if you are going to less common and quite distant destinations, working through a consolidator may save quite a bit of money, especially if you want to fly premium economy or business class.

- **Consolidators are travel agents** who have made arrangements with certain airlines to sell discounted tickets on their flights. Your employer may use one or can recommend one or two; otherwise, go online. The discounts can be modest to substantial, especially on business class tickets, if the consolidator has arrangements with airlines that serve your destination (if not, try another consolidator who does have such arrangements). Even if you are using a consolidator, we recommend that you first go on the internet to determine which airlines fly to your destination in a reasonable amount of time, which itineraries appeal to you, and what the costs are. Note, buying these tickets can be a nightmare. For example, the price will be different at different times of the year, how far in advance and which day of the week you buy the ticket, which days you want to fly, and the convenience of the itinerary. Note also that flights may not operate every day to some locations, so an itinerary that comes up for one set of dates might not come up for another set. Some websites offer such information; others do not, leaving you with several hours of homework trying different days, etc. Also, check the websites of the airlines you think you want to use for other information.

- **Whether you fly premium economy or business class may depend on several factors.** Premium economy provides a bigger seat, more leg room, and often an upgraded menu. It is close to what first class used to be. Business class is now the upscale class with more of everything, including lay-flat seats for ease of sleeping. First, are you allowed by your employer and the granting agency supporting your travel to travel in these classes? Second, it will cost more, perhaps a thousand dollars for premium economy trans-oceanic but several thousand dollars more, maybe three to five times more, even if you get a great price for business class. If the cost comes out of the project grant, there may be better uses for that money. Third, sleeping on flat-bed seats in business class is a great advantage, especially on very long trips, if you are physically challenged for economy class (it does not take much), or just getting up there in age. Plus the drinks, food, wine, and pampering can be exceptional. But recognize that on most trips to Western Europe, for example, you will only get three to four hours of sleep between dinner and breakfast, and the extra fare may not be worth it; premium economy might be a better choice. But you'll get many more hours to sleep on a trans-pacific trip, and that may be worth it. Also, go online for reviews of premium economy and business class service, amenities, and so forth – they can be much better on some than other airlines.

- **Plan for jet lag.** It seems to us to be much more difficult to fly east than west, but even so plan to arrive one or two days before you expect to start work. You can prepare in advance by progressively shifting your eating and sleeping hours to match that of your destination for several days before your trip (a day for each three-to-four hours of time difference at your destination). It is also sometimes recommended to get on the trans-oceanic flight, immediately go to sleep, and do not drink alcohol. This actually may be a blessing if you are flying economy. But the drink and food in business class is part of the experience, in which case this "cure" may be worse than the "disease" for some people. Talk to your doctor before taking sleeping pills, and do not drink alcohol with them. But instead of these general one-size-fits-(maybe)-all strategies, you can get the *Timeshifter.com* app and purchase a scientifically based individual plan to minimize jetlag for your specific trip.

- **Go on the websites of airlines that look promising and determine which airlines belong to the same alliance (e.g., Star Alliance, etc.).** Your fare likely will be cheaper if you stay on airlines that are all in the same alliance for your entire trip.

- **Avoid buying separate tickets for different legs of a single itinerary** (e.g., one ticket to get to New York and a separate ticket from New York to your destination). This can be cheaper, and it is one way to mix airlines that do not belong to the same alliance. But it means you will have to retrieve your luggage, check in, and pass security in mid-trip, and neither airline is responsible for you if your initial flight is late and you miss your connection (i.e., you may forfeit your purchased flight and have to buy it anew at likely last minute prices). Besides, it's nerve racking if your initial flight is late. Note that buying a separate ticket may be much cheaper (called a "mixed fare" or "hacker fare" on some websites) and safer if it is an isolated leg of a trip and you are not connecting with another flight on the same day.

- **Make sure the airline has your cell phone number**, not the consolidator's or other travel agency, to call if there is a delay or cancellation.

- **Pick itineraries with sufficient lengths of layovers** (at least 90 minutes domestically and two hours internationally) and at least two to three hours upon return to your home country's gateway city to allow for passport and customs control. But also be mindful of very long layovers (5 or more hours) on some itineraries.

- **Once you have two to three possible itineraries selected, propose them to the consolidator or travel agent.** Specify if you must fly on certain airlines, how flexible you can be on travel dates, and any other limitations and preferences. You may get a quote for one or more of your identified itineraries but be prepared for quotes for itineraries you did not select. Ask for cheaper fares but be mindful of additional inconvenience in compensation for them.
- **If you are traveling a long distance, you may want to stop on route at another city for some personal sightseeing.** Stopovers cost extra, even if you would have changed planes in that city anyway, and they may complicate piecing together an itinerary and cost considerably more if your extra city is out of the way or only served by airlines that are not in the same alliance as your main carriers.
- **Once you are ready to buy a ticket, there are other choices.** For example, if two or more of you who do not have the same last name are flying together, tell the agent you want to be "linked" in the airline's reservation system. Then, if they want to move your seat assignment they are more likely to keep you together. However, this is not guaranteed. Even in business class, there are unadvertised categories within business class based on the fare you paid, and if you got a bargain fare you may be moved without your knowledge and not be together. If they change the aircraft from the one they used when you originally booked, be especially vigilant. Check your seat assignments at the airport when you check in to see that you have the original seat assignments on your boarding passes.
- **Make sure your name on your airline reservation is exactly – in every detail – the same as on your passport.** If not, you can be denied boarding, and this may take hours to fix if at all.
- **There are some preferences in seat assignments** (even a website – *SeatGuru.com* – that will provide seating charts for each type of aircraft for each airline and will tell you which are the more desirable seats and which to avoid). Although the bulkhead seats may offer extra room, some people find them less convenient for entertainment and meals, there is no storage under the seat in front of you, plus they are located near the rest rooms and galley/bar which are high traffic areas and people may be standing in the aisle next to your seat. Seats in the back of a section can run the same risks plus they may not have your entree of choice by the time they get back to you. So take seats in the middle or slightly in front of the middle of a section.
- **If two of you are flying business class and want to sit together, be sure to investigate the seating configuration on the aircraft you will be using to go across the ocean** (again see *SeatGuru.com*). On some planes, each seat in business class is isolated from every other seat – you literally cannot see or talk to anyone else in the section. This is great for solo travelers and for sleeping but bad for

conversation and working together. On some aircraft they have adjacent seats with a lift-up partition between them that gives you the best of both options.

- **Check the baggage limits.** If you will fly separate legs during your trip, check the baggage limits on those airlines and for those countries – they may not allow as many or as heavy bags as your trans-oceanic flights, and the extra charges can be substantial.

- **If headphones are important to you**, check with the airline regarding whether and which headphones will be provided in your class of service or whether there is a charge for them. If you want serious sound-blocking headphones, better bring them with you if you are in economy (they are provided in most business class cabins).

- **Consider whether you want to stay in a lounge between flights in cities where you will change planes.** Airline lounges offer a much quieter, comfortable environment to rest and perhaps work between flights, especially if you have a substantial amount of time between connections. They also provide complementary food and drinks. You can join an airline's lounge or one operated independently for a rather hefty annual fee, and some executive credit cards provide access to certain lounges (see below). On the other hand, you can get into some lounges for a single-use, one-day fee (check out *LoungeBuddy.com*).

Checking In and En Route

- **Pack lightly, and mind the weight of your suitcase(s).** Some veterans advise rolling underwear and even shirts, pants, and other garments and packing them in airtight plastic bags – they claim they take less room and don't wrinkle as much. Pack these rolls vertically in your suitcase so when you open your suitcase many different types of items are readily visible. Place medications in original pharmacy containers in a carry-on along with a minimum toiletry kit for refreshment and a change of underwear in case you miss a connection. Have a passport holder that screens out electronic hackers and that is easily accessible along with a pen during the flight to fill out arrival documents. Take a blindfold, ear plugs, a neck pillow, and socks or slippers (with hard soles or disposable, because lavatory floors can get wet) if you want these. All these items are provided in business class.

- **Check your seat assignments at check-in.** Some airlines, even those in the same alliance, may not provide you with a boarding card with a seat assignment for a connecting flight. You can ask if seats are assigned (the agent may or may not be able to tell you); if not, see if the agent can assign you seats. Then, when you reach your connecting city, immediately go to the service/connections desk of your connecting airline to get a boarding pass.

- **Obtain an app for the airlines you will be traveling on.** This may be a frequent flier app, which has some advantages. As soon as you know a flight is delayed and you risk not making your connection, call the airline with which you have booked your ticket. Those agents especially at the frequent flier desks are quite good at this sort of thing and are used to handling frequent flyer customers. Waiting in line at the airport can take a long time, alternatives may no longer be available by the time you get to the desk, and agents at boarding desks have less discretion than the frequent flier personnel. It may be helpful for you to have previously identified a later flight on the same or on another airline so you know what your alternatives are if you miss a connection and can ask about them.

- **Know your rights and options if a flight is delayed, canceled, or overbooked and you miss a connection.** The airline is not responsible if the delay is weather related, but they can still be helpful arranging an alternative. Also, if necessary, airlines often can get "distressed passenger rates" at airport hotels and may offer a food voucher at airport restaurants. But if the delay is their fault, they are responsible for getting you to your destination within four hours of your scheduled arrival on international itineraries. They will try to accommodate you in the least expensive way possible, but if you know what you are entitled to, you can be very calm, professional, polite, and persistent as you remind them of the rules. Although it may be the last option offered, if possible (but it may take some persistence) you want cash compensation not vouchers, which can only be used on that airline, often within a year or so.

- **If you are going to make this trip frequently, you may consider getting a high-level credit card designed for frequent travelers.** For about $500 per year, you can get numerous perks, including basic trip cancellation insurance (which would likely cost more for one trip than the cost of the card), other insurance for lost bags and the costs produced by delays, and entrance to airline lounges at major airports. You can get all of these benefits except the lounge entrances for much less on some cards, and as noted above, you can often buy one-time entrance to some lounges. But a 1-year enrollment in a single airline lounge will cost more than the $500 credit card. A business class ticket gets you into that airline's lounge before the business class flight, but not after arrival if your next leg is economy class.

- **Exercise while on board.** On trans-oceanic or other long flights, consider wearing compression socks and doing a few simple exercises during the flight to minimize the likelihood of blood clots in the legs. Some simple exercises, such as pushing your feet back and forth, can be done occasionally at your seat, or bounce up and down on your toes while standing in the aisle.

At Your Destination

- **Local passport control and customs may be a breeze (usually) or onerous (rare but it happens in some countries).** Many officials do not speak more English than the necessary questions they routinely ask. If the formalities look a little foreboding and there are two of you, send one of you first and have that person wait after clearance as near to the passport booth as allowed to see that the second of you gets through without a hitch. If there is a problem, the cleared person may be able to get help for the detained one or they can go through customs to get to your host for help, if available, but once through customs, you cannot get back into the area.
- **Know the customs limits** and what materials you are not allowed to bring into the country and not allowed to take with you when you leave.
- **Go to the ATM** at your destination for a cash advance. You will likely need local currency to pay for airport-to-town ground transportation and living expenses. Obtain local cash at your arrival airport or at bank ATMs. Make sure no one is watching your transaction, and put your hands over and under your card when you enter it into the machine and over the digit pad when you enter your pin number. It is best to use an ATM at a bank that is open at the time in case your card is not returned by the ATM. Some cards or ATMs will only give you a limited amount of cash in one transaction; if so, finish that transaction and start another, except if the limit applies to a daily cash allowance. Converting cash at a bank is usually more expensive (mind both the exchange rate plus the cost of the transaction).
- **Chances are you will have a local contact with whom you will be working who can help you make local arrangements.** Perhaps they will meet your plane (be sure to have an email and phone number in case your plane is delayed or canceled) and take you into town. Otherwise, go online to the airport's website and investigate ground transportation. If there is more than one of you, a taxi may be cheaper and more convenient than the usual airport-to-town bus or train-plus-taxi to your hotel.
- **Investigate housing options.** Hotels are convenient and set up for travelers, and if you are only going to stay two to four nights they might be simpler. Otherwise, your local host may be able to recommend an alternative or you may want to go online to Airbnb, VRBO, or HomeAway to rent an apartment, which is likely to be much roomier and cheaper. This is especially helpful if two or more of you are traveling together and can rent a two or three bedroom

apartment, which will offer room to work at night, a living room in which to relax, and a kitchen for breakfast and even dinner if you want to food shop and cook (but local restaurants are part of the culture and experience). You will need to know where you will be working so you can get an apartment close to that location and have your host vouch for the neighborhood.

- **If you will not return to the country of your destination, you may unload local currency** at the end of your trip by paying at least part of your lodging bill or your last meal with the excess cash.

Returning Home

- **Familiarize yourself with your country's customs limitations.** US customs are much more relaxed than they used to be, but nearly any food product is a no-no.
- **Note that many airports have become shopping malls or even tourist attractions** (e.g., Singapore, the new Istanbul airport). The duty-free stores may or may not be cheaper than the same item at home or in the town of your destination, so it's best to know the home-town cost of items you want to purchase. Buying liquor or wine in a duty-free shop at the airport on the way home is common (but not necessarily less expensive), but the bottles must be put into your suitcase after arriving in the United States before rechecking your bag for a connecting domestic flight.

We told you international travel was more complicated.

Illustrations

In nearly three decades of international travel, we have learned many of the lessons described above the hard way. While the list of advice seems long, perhaps unduly detailed, and even overly cautious, we've experienced most of it. Some of what we illustrate below can be avoided by having a smart phone and taking the precautions we have advised above. But stuff happens!

Getting There

While you may not have control over certain events, you should be prepared for what to do in case you miss a connection or some other unforeseen interruption occurs. Don't buy separate tickets for different legs of a continuous journey, make sure your airlines have your phone number not an agent's, have the app for each airline on which you will travel, and know alternative flights and even routings if you miss a connection.

We were connecting through a major Western European airport. When we got off our arriving flight, the airport was in chaos. People were everywhere, lined up at this or another counter. It turned out there was a work stoppage at the airport, including mostly baggage handlers and the like. Not much was working; flights were not leaving, zip. Our airline said they did not know when flights would leave, but probably not today, maybe tomorrow, this week, who knows!

We went to a bookstore in the airport and opened a travel book to that city and found a phone number of a service that booked hotels near the airport. We called and got accommodations ahead of the crowd, and there was a convenient shuttle bus from the airport to the hotel.

Great. Now, what about luggage? Someone said it was downstairs near baggage claim. Seems reasonable, almost obvious. But coming down the stairs we saw an enormous mound of luggage, bags laying upon bags, in the middle of the floor with people crawling over them to find their bags. Is this luggage from our incoming flight? Other people seemed to think so. This was chaos! We poked through the people first here, there, and then again further around the perimeter of the crowd, trying to see if our bags were visible and reachable. We were lucky, they were near the edge of the pile.

We got on the shuttle bus to the hotel and then started calling the airline to try to reserve space on some connection to our destination. We got out the next day, but we were warned to get to the airport early. Indeed, you could barely get in the doors, and then finding the right check-in counter was difficult trying to wade through hordes of people standing in line with baggage. It took several hours, but we made it.

Delayed Again and Again

We were at the Pittsburgh airport bound for Western Asia. To save money and get the flights we wanted, we bought separate tickets from Pittsburgh to JFK airport in New York, and then another set of tickets on Turkish Air from JFK to our destination via Istanbul. We were aware of the possible problems with this arrangement. We would have to retrieve our baggage at JFK, change terminals, check in with Turkish Air, and go through security before the flight out of JFK. So, we made sure we had at least a four-hour layover at JFK to accomplish all this, and we had investigated the location and route within JFK from baggage claim to the Turkish Air check-in. We were prepared, sort of.

But we did not plan on weather in the New York area, and we did not know that one runway at JFK was being repaired and out of service. The departure time for our flight from Pittsburgh to JFK was delayed 30 minutes. No problem, we have four hours. Then, it was delayed another 30 minutes. Still no problem. And another 30 minutes. Now, we are getting nervous.

Chris noticed that the airline had not emailed or texted us about these delays. She called our consolidator who sold us the tickets. The airline had notified them, but our representative was not on duty so no one notified us; only the gate agent in Pittsburgh posted the delays at the check-in counter.

Chris called the Pittsburgh-JFK airline. They said there was weather in NYC and one runway was closed. Landing priorities were being given to incoming transatlantic flights, and domestic flights were having to wait. How long? They did not know, but they thought we would probably make our connection.

Chris called Turkish Air. Perhaps their flight was delayed arriving at JFK and thus our flight will be delayed leaving JFK. Nope – on-time departure! If we missed this flight, is there space on tomorrow's flight? No, not at the fare we paid! Yes, they had seats but not for us business class folks who got a bargain fare. Besides, there is no flight from Istanbul to our final destination the next day. These flights do not operate every day. We could go the day after that, but then we would have to spend two nights in JFK/NYC.

Maybe we should go home and come back to the Pittsburgh airport in two days. Well, there was no space on the Pittsburgh-JFK flight that day. We could go to Boston and then JFK. This is a mess. We decided to wait a bit longer to see when our flight to JFK would actually leave and whether we would certainly miss the connection. This was getting tense.

We left Pittsburgh for JFK after a three-hour delay. We were nervous about whether our plane would have to circle around before landing, eating up more time. We were not the only ones in a bind. There were several people who had already missed their connection to Europe for a cruise. They seemed in worse shape than we were. We landed at JFK with about 45 minutes to go. We ran to the baggage claim, which seemed like a mile away from the arrival gate. Once there, we waited for about 20 minutes for our bags to arrive. We ran with baggage to the train between terminals. We were down to about 10 minutes. When we got off the train we could see below the massive crowd waiting to check into Turkish Air and an even bigger jam-up trying to get through security.

Fortunately, we were flying business class, and there was no one waiting in that line to check-in. That process was fast; security not so much. We did make it, but only by a few minutes. Never buy separate tickets! Know how to contact the airlines you are flying on. But business class was worth it, and we had not even gotten on the plane yet.

Foreign Passport Control Issues

In some countries where there is substantial government control, entering the country can be a little intimidating. If you have all the required documents in order, it should not be a problem. But it can be a little anxiety provoking.

After landing, we walked with the other passengers into the terminal building. We came to an open area, and the other passengers were ahead of us, lined up in front of several little booths. The booths had solid walls and a door, except on one side where each passenger, one at a time, stepped before an open counter behind which a customs official sat. There were perhaps ten people in front of us at each booth. The lines moved very slowly. People sometimes spent a long time getting their passports and visas checked.

Suddenly, two officials, armed with handguns, emerged from behind us and moved to one of the booths. A man was pleading his case to the custom official. The two security officials went up to him, each grabbed one of his arms, and led him off to the side and into a room. Bob looked at Chris with some apprehension, "Mmmmmm!"

"I think we need a plan here," Bob whispered to Chris in the relative silence of the crowd. "If one of us gets pulled out of line, there is no guarantee anyone will speak English, and we will not know what the problem is. Also, there is no one here to help us, and our local colleagues cannot get into this area. So, you go first, Chris. If there is a problem, turn toward me and tell me whatever you know about it. If they lead you away, I'll try to come with you. If they will not let me, I'll try to go through passport control. I might have the same problem, but if not, I'll go through the baggage and customs area and inform our local colleagues who are supposed to meet us. If you get through OK, wait on the other side as close as they will let you to see if I get through. If I don't, then you get our colleagues."

The experience in the little booth was similar to Chris's first impression (Chapter 4). It all seemed unnecessarily intimidating, but we both made it without a hitch. Bob experienced this kind of treatment years before when he drove from Western Germany through Eastern Germany to Berlin. It seems like the intent is to convey who is in charge – the officials are, and they ARE in charge.

On to baggage claim and customs control. Chris goes into one line and Bob into another. Suitcases are to be put on a long table and then opened. The official paws through the suitcase, underwear and all. Bob is dismissed rather quickly, not so for Chris.

Her official asks her something in the local language. Chris looks bewildered and slowly shakes her head and says, "I don't understand." The official is insistent and tries again. Chris shrugs. He is getting irritated. He rummages a bit more in her suitcase and utters something else. Chris says, "I don't understand." Finally the official apparently gives up, and shouts something that sounds like "Id-YOT," and he waves his arm toward the door. Chris packs up quickly, and hurries through the door.

We meet our colleagues on the other side, having navigated the local immigration gauntlet. But immediately after the "hello's" and "how are you's" Chris bursts out, "he called me an IDIOT!" Chris is otherwise the consummate diplomat, but no one calls her an idiot and lives.

"What do you mean?" our colleague asks.

"He went through my suitcase, asked me something three times. I tried to explain I did not understand. He got mad, waved his arm at me, and called me "IDIOT."

We all laughed, except Chris, of course. Several years later, Chris repeats this story to our colleagues at a social occasion. Our colleagues listen, and then suddenly one bursts out laughing. "He didn't call you an idiot. 'Id-YOT' in our language means 'exit' or 'pass through.'" Even Chris laughed in relief.

We went to this same country numerous times over the next two decades. Things changed dramatically over that time as the economy improved. The airport terminal building was remodeled, and customs and baggage check were essentially eliminated unless you had something to declare. They even built new passport control booths, but they kept the same design of bright lights, high shelf, small opening in the glass panel, and no smiling or conversation. The government was still in charge.

From Airport to Town

Ideally, you want to have your local colleagues meet you at the airport, or at least send a car and driver to pick you up.

If not, research the airport website for ground transportation to town. There will probably be several alternatives, some of which can be arranged in advance. Otherwise, transporting foreigners, especially Americans, from the airport to town can be a dicey enterprise. There may be taxi cabs, with or without identification on the cars; there may be "officially credentialed" drivers or limousine services; and there will be plenty of "freelancers" looking to make some money. The fare can range from modest to exorbitant, and your safety – who knows. The "officials," such as police or tourist guides, may be trustworthy and maybe not – it can vary from country to country. Most of the time, nothing untoward happens, but it's not guaranteed. Heck, we've been ripped off in Rome and Athens, let alone more remote corners of the world.

The drive into town may be enlightening. We often arrived at night in some countries. It was usually hot. In many countries, the poverty was pervasive. Many people were outside eating, drinking, talking; children were playing; and dogs ran freely. The places of business were usually open air and very basic. Houses might be made out of scrap, including large pieces of cardboard, sheets of metal, some walls of wood. Other people may live in the hills off to the side of the road in even more basic dwellings. The people seemed happy, but incredibly impoverished to American eyes.

Mind the Weather

Bad weather occurs all over the world, and it can be especially disrupting when you are scheduled to fly. The following illustrations occurred years before modern technology (i.e., smart phones, apps), but they illustrate why you now should register with the STEP service of the US State Department as advised above to warn you of serious weather, civil disturbances, and the like. Also, get the app of the airlines you are using so you can be notified of delays, flight changes, and so forth. This is what you do now, but this was then.

We arrived at the airport in plenty of time for our flight to the United States. Everything seemed quite normal, and there were plenty of other people checking in for their flights. We went to an airline lounge to have a snack before getting on the plane.

After about 30 minutes, Chris observed, "There used to be quite a few people in this lounge when we arrived, but it's thinning out now."

"Maybe several flights depart now and people left to get on their planes," guessed Bob.

"Maybe," Chris responded but skeptically. Perhaps 10–15 more minutes passed, and now there were only two other people left in the lounge. So we wandered over to the TV listing all the arriving and departing flights. All the flights were flashing "Canceled." We hurried over to the lady at the lounge reception desk, "Why are all the flights canceled?"

"Meetch! Meetch! Time to go," she said with some urgency.

"What's 'Meetch'?" Bob asked.

"Hurricane.... storm," she blurted frantically. The hurricane had been named Mitch. "Airlines have bus, take passengers to town. Get baggage at baggage claim. Bus outside there."

"Hurricane?" Chris asked Bob. "Did you know about this?"

"Nope. We've been busy. Neither of us have been monitoring world events or the weather the last few days. But why did the person checking us in for our flight not say anything or someone announce this in the lounge?" Our luggage was indeed at baggage claim, and outside numerous other international passengers were lined up waiting for the bus under a shelter from the rain, which now was getting heavy. The mood was jovial, which masked a good deal of uncertainty and nervousness.

"You ever been in a hurricane?" Bob asked Chris.

"Nope. I could skip it."

"Where are they going to take us, does anyone know?" someone asked the crowd.

"Some hotel in town," the "crowd" answered. Town was more than 20 miles away, and the rain was getting heavy, very heavy. A large bus finally pulled up, and all of us helped to load our luggage into the compartment in the rain. We were drenched, but glad to be on the bus.

The bus pulled out of the airport and onto the highway toward town. The rain was now blinding, and the bus could only go about 30–40 kilometers per hour. The local people were scrambling along the side of the road, trying to move articles toward shelters and getting out of the rain themselves. Puddles gave way to ponds on the roadway, and occasionally the bus would fishtail a bit through the water. People emerged out of the foliage along the road looking for shelter of any kind. Women carrying babies were running through knee-deep water. They may have lived in the hills along the side of the roadway. Later, we learned that the mud on some hillsides gave way in places and slid into the valleys, and some people were killed. Occasionally the water on the road must have been a foot or more deep, and we all grabbed an armrest or each other each time the bus fishtailed. It took more than an hour to go the 20 miles to town and a hotel. There was no electricity and no elevator to our rooms on the fourth floor, but somehow, it looked so good. We left the next day, uneventfully.

Actually, we were lucky. We should have been more aware of the weather. Mitch hit land on the Eastern coast of Honduras and slowly moved across the mountains and northward through Nicaragua and El Salvador. While it lost some velocity, its slow pace permitted horrendous amounts of rain and flooding. Overall, 11,000 people were killed in Central America, and countless numbers of homes were destroyed and people rendered homeless (http://news.bbc.co.uk/2/hi/americas/202395.stm). We WERE lucky – the human tragedy was in the rural areas and hills. We only saw the rain and some flooding, but that was enough.

Flight 34 Departing

We had landed in San Francisco to connect to a flight to Hong Kong. We had a two and a half hour wait, so we walked around the airport and looked at the shops. Then, the airport public address announced:

"United Airlines Flight 34 to Hong Kong is now boarding at Gate A 12. All aboard please."

"That's our flight," noted Bob, "but it's way too early to board. We don't leave for another two hours or so."

"But if that's our flight," cautioned Chris, "maybe we should check it out. Did we miss a time change or have the wrong departure time?" The gate was a bit of a hike.

"United Airlines Flight 34 to Hong Kong is now in its final boarding stage. All passengers should be on board."

Our pace quickened to a run. We made it to the gate just in time and scrambled to our seats. After we stored our carry-ons in the overhead compartment, put on our slippers, and sank into our seats, Bob murmured to Chris, "I've never been on a flight that left early like this. We must have gotten the time change wrong. Oh, well. Is it cocktail hour yet?"

We arrived in Hong Kong in the early evening at their new airport some distance out of town. But the place was nearly empty – no people. After collecting our bags, we went to the taxi stand to get a ride to our hotel in town. There was one taxi available, no others in sight. We got into the car, and off he sped. We flew to town. It started to rain.

We checked into the small hotel that our host had arranged for us. No one was in the lobby except one desk clerk who was fixated on the small television screen on the wall in the corner.

"What's going on?" Chris asked.

"Typhoooon! Typhooooon!"

"Typhoon, as in storm?" Bob asked.

"Typhoooon!"

"OK. Guess we hunker down for the night and maybe tomorrow. That must be why the plane left early to get here before the storm hit." We later learned we were the last car over the bridge from the airport to town. It rained the next day but nothing serious. Hong Kong took a glancing blow. Mind the weather; get the STEP and airline apps.

Events on Board

Sometimes strange things happen on flights.

Bob had fallen asleep on an overnight trans-oceanic flight, but Chris does not sleep so easily on flights. Then, Bob woke up to go to the restroom. He started to get out of his seat when he noticed a body lying face down on the floor in the aisle.

"What the hell is going on here?" he asked Chris. "Is this person OK? I mean she – I guess it's a she – obviously is not OK. How bad is this?"

"They carried her back here and laid her down there, and she hasn't moved," Chris reported. "They called for a doctor, and a man said he knew a doctor in business class, who then examined her with the help of the airplane's medical equipment. About all he could determine was that she was alive. She is probably sick with something, but they all decided to just lay her down where she was, and she was right here."

The middle seats across from us were vacant. We suggested she might be better laying across those seats. Some other passengers helped her onto those seats. Later, she suddenly lurched forward and vomited.

"Whoooah. Sick all right," Bob declared the obvious. "What a mess." The poor flight attendants had to clean it up the best they could. An ambulance was waiting for her when we landed.

Told you – stuff happens, this time to someone else, sort of.

I Need a Shot

Health concerns, especially unusual medications, can produce issues while flying and when going through security. Be sure to have all unusual medications in their original pharmacy bottles and perhaps notes from your physician authorizing you to have certain medications.

Chris has suffered migraine headaches nearly all her life, as many as two or three a week. A physician at the University hospital specialized in creative treatments for migraine sufferers, but when all else failed and the pain was sufficiently intense, he had prescribed an injection with disposable syringes that nevertheless had to be filled with the medicine before administering it. We are in the middle of a trans-oceanic flight to Latin America, with a young woman on the window, Chris in the middle, and Bob on the aisle.

"I've got to take a shot," Chris moaned to Bob. "I'm shaking. Can you fill the syringe for me? I'm not steady enough right now to do it."

"I'll try," answered Bob. "I've never done this before, but I've seen it done many times." Chris handed him the syringe and the small bottle of medicine. He held the bottle upside down at his eye level, and slowly penetrated the rubber cap with the needle. Then, he drew the plunger slowly out to fill the syringe. He pushed the plunger in to get rid of any air bubbles. At this point, a flight attendant walked by. She looked, paused ever so slightly, then continued down the aisle. "Suppose we'll get arrested upon landing for doing drugs," Bob asked rhetorically? When the syringe was full, he handed it to Chris.

"Thanks." Chris jabbed the needle into her solar plexus and emptied the syringe's contents. Then, she put the used syringe back in its container and into her purse.

Our seatmate appeared to barely notice this drama, but Chris felt she should explain, at least a little. "Migraines," she started to explain to our seat mate.

"It's OK," the young woman said. "I gathered it was something like that. I'm a neurophysiologist."

We were not arrested upon landing.

But departing that country was a different story. Chris's carry-on bag tripped the scanner going through security, and she had to open it up for the agent. She had perhaps ten syringes and vials of medicine. The agent insisted she could only take one syringe and one vial with her on the flight to the United States.

"Why should they care what I take out of the country?" Chris complained to Bob. "I understand the United States might be nervous about that, but what difference does it make to them here? Oh well, one would be enough to get me home if I need it. Would you fill another syringe for me?"

"Sure."

7 Living Abroad

General Lessons Learned

Plan on numerous trips to your project site before and during the project. An initial visit to another country to explore possibilities for a new project may last only a few days or a week or more, depending upon how much advanced planning has already occurred. Later trips to plan all the details of a complex project can require visits of two weeks or more. Further, a major project involving many participants, numerous measurements, and multiple sites may require someone from your home country to live on-site for months or even years. Indeed, we were advised by an administrator at a major commercial concern that they would never operate a substantial project in another country without sending a representative of the company to live there during all phases of the project.

Numerous trips per year for weeks at a time or a representative living on site may strike some granting agencies as excessive, in part because travel funds are typically scrutinized carefully on research budgets for domestic projects. You will undoubtedly need to justify completely the costs of periodic trips, especially those lasting two or more weeks, or a resident representative's living expenses. Although your international colleagues may vary in how communicative they are with you – the internet and Skype now make communications much easier and cheaper than previously – in our experience the trips were absolutely worth their expense, especially to a site where a very large, comprehensive, empirical project was being conducted. The trips also motivate your international colleagues to get things done before you arrive, and you may discover small and occasionally large details get decided while you are away at home.

Investigate living options. Where you stay depends on how often and long you will stay per trip and the economic circumstances of the location of your project. For example, one or two people making a short trip of a few days would probably be best to stay in a hotel. This is easy in a large, modern city, but if you are going to a very low-resourced city or a more rural location, you will need the advice of your local colleagues. You will want to be near where you will work, so their advice about location, safety, and amenities is always advisable. Find out about how local transportation needs will be met, including airport to town, living accommodation to work site, visiting several organizations in town, etc.

For longer stays in cities, explore renting an apartment, either privately or perhaps through Airbnb, Vrbo, or HomeAway, but again check with your local colleagues about location and even have them visit the apartment and perhaps arrange to rent it on your behalf. If it is a private arrangement, explore how you will get a receipt for the rent that will pass your institution's accounting requirements. One strategy is to have a local agency involved in the project rent it for you, you reimburse the agency, and the agency gives you a receipt. Apartments are much more spacious than hotel rooms, can have two or three bedrooms to accommodate a slightly larger group, provide living space and cooking options, and are much easier for two or more people to debrief and work together after hours. And they may be much cheaper if two or three people are traveling together and especially if someone in your group is willing to cook occasionally. Experiencing local restaurants provides a window on one aspect of the culture, but so does local food shopping and cooking. If these arrangements can be made smoothly, this can be the fun part of such projects.

Illustrations

Over the years we have worked in several countries around the world. Our living experiences varied with the particular country and the nature of the project we were conducting there.

Exploring Alternatives

Chris, Bob, and a work colleague visited Russia a few times to explore political factors affecting the institutions and to meet officials who might influence the quality of care for vulnerable children (see Chapter 2). These trips were initially made in the 1990s when Russia was under substantial economic distress. Quality hotels were not on the country's list of priorities.

Early hotels. On an early trip we stayed in a very basic, small hotel, not accustomed to having many guests. There was no lobby, only a desk, and occasionally there was someone available to assist guests. The rooms were very small. Bob had a divan for a bed that occupied half the length and width of the room. At its end, there was a very small desk, a radiator, and a window. It was winter, and thank goodness the radiator worked, so well that the window had to be opened to maintain a livable temperature in the room. A tiny bathroom had a toilet and shower, but Bob got a mild electric shock when he touched the plumbing. The room was "hot" in more ways than one. Chris and our colleague's room was only slightly bigger and not better. They had to put a long blanket over the window to keep out the wind.

When Bob and Chris started to make regular trips, they needed to explore a better hotel that was nearer to the agency in which they would meet and work. There was an up-scale hotel about a half-mile from the project's location. The rooms were certainly larger than the previous hotel's rooms, nicely appointed, and the water, but not the plumbing, was "hot." But they rented for about $250 per night even in the late 1990s. We noticed that restaurants were also quite pricey, at least the ones that we felt comfortable eating at and had English menus (now one can get smart-phone apps that translate menus – we recommend having one).

"If we stayed here," calculated Bob, "it's going to cost $500 per night for two rooms plus at least $150 for food for the two of us. For a stay of maybe ten nights, we're talking $6,500 plus air fare per trip. If we are going to pursue this project, we need to find a cheaper way to visit than this."

"I agree," said Chris. "Let's ask our local colleagues what our options are."

The first local apartment. Part of the next day's meeting with our colleagues focused on where to stay near the agency.

"Oh," said Nikita, "you could probably rent someone's apartment around here for much less than the hotel. Some people actually move out and live with relatives or somewhere and rent out their apartment. After the renters leave, they move back in. It is very much worthwhile for them to get the extra money. Technically, they are not supposed to rent out their apartment. Apartments were actually owned by the government and were allocated to people under the Soviet system, but now the building still belongs to the government but the apartment itself sort of belongs to the people to whom it was originally allocated."

"So, renting an apartment is actually illegal?" summarized Chris. "Is there any danger to us in doing this?"

"No, no danger to you," replied Yuri. "Some people actually advertise their apartment in the English newspaper, and some richer people buy them and rent them out as a sort of business."

"I'll find an apartment for your next trip," offered Nikita. "I'll go look at it for you before we agree to rent it." Nikita found a two-bedroom apartment for the next visit that was quite close to the project agency for $600 for 10 days. It was a fifth-floor walkup that took several trips up the stairs with luggage to move into. It consisted of a long hallway; off the hall was a living room, two bedrooms, an interior bathroom, a dining room, and kitchen. At the end of the hallway was literally a "water closet," a toilet on a raised platform that was missing parts of a few floorboards in a closet-sized room with one small incandescent bulb illuminating the chamber. It was an interesting sight to enter the apartment and look down a dark hall to see this dimly lit toilet at the end. Chris was a little concerned about the holes in the floor and what manner of wildlife might emerge through them, so we kept the door closed even though that was not likely to be effective; besides, it smelled a bit in there.

We moved in. It was closing in on winter, and the apartment was a little chilly. These apartment buildings are all heated and supplied with hot water for heat and bathing and electricity by a central facility that serves perhaps 10–12 city blocks. They turn on the heat according to a schedule more or less regardless of how cold it is outside. Apparently, it was not time yet for heat. The water system is not the best. Water in the bath and kitchen runs different shades of brown from one day to the next, so bottled water is a necessity for drinking and cooking. Chris wondered if she would get stained if she showered in such brown water. Bob did not think so but suggested it would be a quick tan if she did.

The next morning Bob showered, and then Chris braved it. Bob had to find an electrical outlet someplace elsewhere in the apartment to dry his hair while Chris was in the bathroom. He found an outlet in the kitchen and started to blow-dry his hair. Suddenly, the hair drier stopped, punctuated by a scream from the bathroom.

"Help, help. There's no light in here. It's pitch dark," Chris yelled from the bathroom.

"OK, OK, I'll get the flashlight, if I can find it. Hold on," Bob pleaded. He found the flashlight, opened the bathroom door a crack, and rolled the lighted flashlight on the floor into the bathroom.

"Thank you" came Chris's response. "What the heck happened?" she shouted through the closed door.

Well, I'm not sure, but I think we – or the building or the whole city – blew a fuse or had a breakdown," speculated Bob. "I'll go to the kitchen where I can see through the window if some of the other apartments in the building have lights.

"No, they have electricity," Bob shouted to Chris who was still in the bathroom. "It's not the city or something, it's us. We must have blown a fuse. I think the refrigerator went on while I was using my hair drier and that was too much. I'll look for a fuse box."

Bob went room to room and down the hall looking for a fuse box, even outside the apartment entrance door and down the five flights of stairs to the entrance way. No luck. Then, he noticed that there was a ladder just inside the apartment's front door, and there on the wall up about 10 feet was a black box that might pass for a fuse box. Perhaps the ladder suggested that this was not an infrequent event! He climbed the ladder and opened the door of the box. Yup, two fuses, but which was the blown one – it was too dark to tell. So, he unscrewed one fuse, and in 20 minutes we were ready to go to the agency to meet with our colleagues and show them the fuse.

"Oh, 'prokpah,'" exclaimed Nikita in Russian. "It happens all the time. I'll get you another one. It should be OK." And it was.

After that, the first thing we did when we arrived at an apartment was to look for the fuse box. It was in a different place in each apartment – up in the corner near the door or behind a picture that was hung over it in the hall, very creative.

Buy an Apartment?

Nikita arranged for an apartment for us on subsequent visits, but this was often at the last minute. Indeed, on one trip we had no assurance before we arrived that an apartment awaited us or where it would be located. Nikita said landlords do not want to commit to a tenant very far in advance in case they get a longer-term renter. So, for us, they will rent for a two-week stay only at the last minute.

"Well, that is understandable from their standpoint," empathized Chris, "but it is a little nerve racking for us. Nikita, do you think you can always get an apartment for us even a few days before we arrive?"

"Well, yes, I think so, but of course it is not guaranteed. Your apartment is actually not very good, but it was the best I could find for the dates you were coming. There are other apartments near here that are better. I saw a couple."

"Well the price was good for this one" said Chris. "Are the others more expensive? We have got to have a two-bedroom apartment. And we really need to be able to walk to here for our meetings every day."

"Maybe a little more expensive, but probably not much – maybe $700–$800 instead of $600," Nikita speculated. "But you could BUY an apartment now and use it for the entire length of the project. Then, at the end of the project you could sell it. It would probably be cheaper overall. They cost about $15,000."

"Can a foreigner buy such an apartment?" asked Bob.

"Probably," answer Nikita, "but there might be some complications. If you gave us the money, we could buy it for you, and maybe we could use it when you are not here."

"That's very interesting," Chris murmured. "We'll have to think about that."

That night, we discussed the idea of buying an apartment.

"Well, for openers, $15,000 is not a lot of money," started Bob. "If one could sell it for about $15,000 at the end of the project, we would be ahead compared to renting, which would be perhaps three times a year at $600 equals $1,800 a year and for at least 5 years is $9,000."

"OK, it could be financially worth it, but I worry about the details," cautioned Chris. "First, where does the $15,000 come from? I don't think the University Research Accounting Office or the government granting agency is going to approve an expenditure listing of '$15,000, apartment purchase in Russia.'"

"Agreed," said Bob. "We would have to come up with that purchase price personally. Then, each trip we would be paying rent to ourselves.

We'd be laundering it through the Russians. If that were discovered our buns would be butter."

"Having our colleagues buy this for us is probably much simpler than if we tried to buy it, but it's a bit risky," demurred Chris. "They could walk off with the apartment and the money at the end of the project. Even if we had a contract with them, what are we going to do, sue them from the United States? Do they even have contracts and lawsuits here?"

"Yes," agreed Bob, "we don't know them well enough yet to have built up the kind of trust that would be necessary for a $15,000 deal. I also don't know if there is homeowner's insurance here; I doubt it. And right now the local economy is very distressed and uncertain, and who knows what it will be in five-plus years."

"Do we agree?" concluded Chris. "We don't do this."

From a financial standpoint, the local economy improved substantially during the course of our projects there, inflation often ran 10%–15% per year, and by 2012, we were told such apartments sold for around $50,000. But the arrangements were very uncertain, and all-in-all, it was good we did not buy an apartment.

A Regular Apartment and the KGB

Over the next few trips, we stayed in different apartments, and we gave each of them a name. One had a very large living room with a grand piano and a candelabra on it. It looked just like Liberace's setup, but we thought labeling this apartment the "Liberace Apartment" was a little culturally inappropriate, so we called it the "Tchaikovsky Apartment."

Eventually, we started to rent from the same landlord. He owned or controlled several apartments in the same building, so if one was taken during our planned stay, we could have one of the others. The building was two blocks from the project agency, two blocks from the American Medical Clinic, and walking distance to a large food market. It even had a little elevator. Judging by the high-end cars parked in the courtyard, this must be close to "up-scale."

The friendly landlord had done all the modernizations and repairs to the apartments himself. Some of these touches were interesting. For example, the hot and cold water plumbing was opposite to the hot on the left and cold on the right, sometimes even opposite to the blue and red color codes on the faucets. The shower stall leaked once, but he fixed things fairly quickly. He and his wife occasionally lived across the hall from one of the apartments we rented – we also rented that one a few times – and they had a very large, old Russian Bear Dog, a black, hairy, and entirely placid dog who was often sitting on the hallway landing between their and our apartment. He or she looked like a Russian bear. We called him "Boris" or "Boristina," we did not know which was appropriate. He or she didn't mind.

As noted above, the buildings were owned by the city, and they were typically in ill repair. The door from the courtyard into the building was made of heavy sheet metal with a combination lock plus a key. Electrical wiring hung loosely from the hallway ceiling, and the pockmarked plaster walls resembled a Jackson Pollak painting of splashed colors. No one seemed to sweep or clean the halls. Occasionally, a notice was placed on the outside of the main entrance door. Of course, we couldn't read it, but sometimes it advised tenants that the heat or water would be out of service for a while, so we would run up to our apartment to try the water. Actually, most of the time we think they announced a tenant's meeting.

The apartments, however, were much nicer than the hallway, being the "property" of the residents. Some had elaborate doors with brass fixtures and even the tenants' name on it. Inside, the doors invariably had three locks — a key lock in the handle, a bolt, and a chain. These did not always work smoothly. We had to rehearse opening them in case we needed to evacuate in a hurry. Security is a priority in Russia.

Inside, the apartments were comfortably furnished in Russian style with carpeting and large, puffy couches covered with ornate fabrics. The wallpaper was quite patterned in Eastern European style. But we noticed a curiosity in the living room of all the apartments.

"What is that thing?" asked Bob pointing to a small box in the corner of the ceiling. A wire from the box ran down the wall and then into the wall near the floor.

"I have no idea," responded Chris. "There has been something like this in each apartment. But there is a little window or something in this one. Do you suppose someone can listen or watch us through this thing?"

"I don't know," said Bob. "Do you think we should cover it over with something? We are certainly not doing anything illegal, political, or even very interesting, although someone might think watching the Americans is kind of novel and entertaining."

"No, but let's go sit in the kitchen. I've never seen something there. We can ask our Russian colleagues tomorrow."

"Oh, yes," said Yuri the next day when asked about these little boxes. "That's the KGB's surveillance system."

"Really?" Bob inquired rather incredulously.

"Yes, they keep watch on everything here," Nikita assured us.

"Can we cover it over?" proposed Chris.

"Oh, no, they might get upset," Nikita warned.

"OK," laughed Yuri. "Enough. It is not the KGB. These were little speakers that broadcast Soviet propaganda, music, and news into every apartment. It was kind of like a political version of your National Public Radio, but it was mandatory. It was piped into every living unit, and you could not turn it off."

"But is it used for anything now?" asked Chris.

"No. Forget it. It doesn't work at all."

Upgrading the Apartments

Essentially all the apartments that we rented were adequate in general, but not in terms of bedding and kitchen equipment. The blankets looked a little "used," and we wondered when they had been cleaned last. The pillows were also a bit thin and decidedly not fresh. The kitchens had only the most basic cooking and eating utensils. There were always at least two vodka glasses, but there could be from one to three, maybe four, of any other kind of glassware and dishes. Pots and pans were quite old and a little charred, and the cutlery – well, the fork bent at the handle when inserted into a piece of meat. This was a commentary either on the quality of the fork, or the meat, or both.

"Look," announced Chris, "if we are going to live in these apartments for maybe two weeks three times per year for 5 years or who knows how long this project will go on, we need to upgrade the amenities a bit. After all, we are paying $600–$800 a visit instead of $5,000–$6,000 for lodging at the hotel."

"And," chimed in Bob, "if we are going to do our own cooking for many meals and perhaps entertain our Russian colleagues, we need kitchen stuff and dishes."

"I say," Chris effused, "we go downtown and buy decent quilts and blankets, pots and pans, cooking utensils, dishes, glassware, and cutlery. We get some cardboard boxes that supplies come in from the agency, and we store this stuff in those boxes at the agency between visits."

"Let's see if the agency would provide a car one afternoon for us to shop downtown at that big indoor mall and bring all this stuff back here," Chris suggested. "It would have all of these things."

Two days later, we indeed had a car and driver and bought all of these furnishings. We got the agency to save boxes for us to pack these things and store them between visits. We decided that we would do a good deal of cooking ourselves, so Bob brought from home all kinds of herbs, spices, cooking wines, and other materials for cooking. Both of us brought foam pillows from home upon which to lay our sweet heads.

We inventoried our things and packed them in numbered boxes at the end of each trip. The boxes were picked up by the agency driver and our Russian colleagues and stored at the agency. Before each subsequent trip, we would tell our colleagues which boxes we would need, and they would have them in the apartment when we arrived. We would be met at the airport, taken someplace to exchange money, go to a grocery store, and head to the apartment. Then, we would unpack, have a vodka or two, and pass out after 30 or so hours of traveling.

"Baaahh Baaahh, Moooo Moooo"

When we first started to visit Russia in the early 2000s and cook in the apartment, we shopped for groceries at little local food stores. Big all-purpose grocery stores were just emerging on the scene, and we were not close to one anyway. There were numerous little food stores even in a single city block, each specialized in something but each one was well stocked with liquor.

We shopped mainly at a little corner market where obviously many locals bought their provisions. It was quite small and always crowded. We had to learn the shopping drill. After several mistakes and reprimands in Russian, we figured out the drill. The tiny store had several different sections defined only by the nature of the items sold in each section. Fruits and vegetables in one section, meat in another, groceries in yet another, and so forth. One selected the items in a single section that one wanted, a clerk packaged them as needed and rang up the cost, which was printed on a cash register receipt. Then, one left the items with the clerk, went to a pay station, paid for the purchases, the receipt was punctured with a spindle, and one took that back to the clerk to retrieve your purchases from that section. Then, on to the next section, one section at a time. After several years, this store moved to having two or three checkout cashiers where one paid for all items at one stop. The technology revolution had arrived.

One day our colleagues took us to an enormous market. It had an outdoor dry goods section with basic clothes, shoes, coats, sweaters, etc. Inside a large arena-like building were numerous food stalls with fruits, vegetables, meat, sausage, fish, bakery items, spices, and so forth. This looked so much better than the local food store, and we decided to do our future shopping here. Our colleagues warned us that we might get robbed or cheated. Out of an abundance of caution, we left valuables in the apartment and took only enough money to cover the cost of the items we intended to purchase. Nothing ever happened.

On our first solo visit to the market, we had no trouble buying fruits and vegetables –they were clearly identifiable to us, and all we had to do is point at what we wanted and raise fingers for how many. Clerks could tell we did not speak Russian, and they all had little calculators and would punch in the cost and show it to us. Sausage was another matter. There would be maybe 20 kinds on display. Since we did not know what to buy, we decided to try the leanest sausage that was the most expensive. That worked, but because we did not know the name of the sausage to ask for it the next time, we repeated the process each time. We also used this strategy at the fish counter where they had maybe five kinds of salmon-looking options. Pick the most expensive. It was good.

The butcher shop was a different experience. It hearkened back to meat markets of old, with sides of animals hung near an enormous wooden chopping block. Some cuts of meat were on display in a case, but Russians cut meat a little differently than we do, and it was some-times difficult to know just which animal the piece in the case came from and which cut it was.

Bob improvised. A bubbly woman waited on us, talked a blue streak in Russian even after she realized we did not understand a bit of it.

Bob pointed at what he thought were lamb chops in the case and said, "Baaahh, baaahh?" in his best lamb talk. Chris immediately turned away and covered her face in embarrassment.

"Da," the clerk answered laughing. Then, she held up five and ten fingers, presumably asking how many chops we wanted. We indicated eight – we'd freeze some. Then, she motioned with her thumb and forefinger how thick. We responded. "Da," she announced. But instead of picking eight chops out of the case, she called on her butcher who brought over a saddle of lamb, put it on the chopping block, and with a big machete-like axe, took four swings at the saddle and then chopped each piece down the middle of the spinal column. Here you go, eight lamb chops. The clerk weighed them and announced the price. Who knows if it's the same price anyone else would pay, but it seemed OK to us. The lamb was very good tasting, but it took quite a bit of scraping and cleaning to get all the bone fragments off the chops before cooking them.

Next, we had not been very pleased with the beef that we bought elsewhere. It had been fairly tough. Veal might be better. There were some light-colored chops in the case. Were they pork or veal? Veal would be wonderful, but pork more likely. So, Bob points at the chops in the case and tries his best pig talk, "Oink, oink, oink???" Chris melts in embarrassment again and turns to look at vegetables at another stall.

The clerk is totally befuddled. She kind of shakes her head, clearly, she doesn't understand. Perhaps Russian pigs do not "oink!" OK, forget pork, lets pantomime veal. Bob tries, "Moooo. Moooo!" The clerk's face brightens, and she laughs. Then, Bob motions by bringing his hands together signaling that it is a little "Moooo." He also points to some pale colored meat in the case and asks, "Moooo?"

"Nyet."

Then, he points at other pieces in the case and hits a winner – "Moooo?" The clerk's face brightens to a big smile.

"Da, da," she says.

Bob goes for broke. He tries, "Filet. Filet?" She understands the French/English word, and then shouts something at the butcher. He runs into the back room and emerges with an entire side of what certainly looks like veal. The clerk points to the filet running her finger down the inside of the carcass. Bob nods approval; it helps to know your veal anatomy. The butcher cuts out the entire filet on one side, she weighs it, and we pay who knows what price per kilogram. Veal filet mignon, tender, delicious. You could never get that in the United States, because the filet is never separated from the loin chops.

The next day with our colleagues, Chris insists – wisely – that they give us a card with the Russian names for various animals, the best salmon, good sausage, and so forth so that Bob does not pantomime 20 questions with the clerk, entertain half the food market, and embarrass Chris. However, the next time we went to the market, that clerk could see us coming 20 feet away and greeted us with a big smile. Either she enjoyed the game, or she made a big profit on the meat. Who knows, but we were happy with the meat, and Chris was happy we could point to the Russian words on the card. Today, technology to the rescue – "Hey, Siri, what's the Russian word for veal?" Apps will translate a menu or even provide live simultaneous auditory translations. Amazing, but not nearly as much fun or entertaining.

We did notice an interesting phenomenon at the market. If one stall did not have something or needed change, that clerk would go to the next stall and just pick the item up, bring it back to her stall, and sell it to us. Further, all the vegetable stalls had basically the same vegetables, and these were the same in stalls on the street corners. And all the stalls in the market were served by the same group of young men who would resupply them by bringing more produce from the back room. We asked our colleagues about this.

"It's the Russian mafia," they said. "They control all the food that comes into the city and gets sold any place in town."

On the street near the market, we observed numerous elderly ladies selling various things, such as shawls, socks, and the like. Some held up strings of dried mushrooms. Russian cuisine has several common in-gredients, including cabbage, potatoes, sour cream, and mushrooms. We suspected that these ladies raised the mushrooms, strung them on thin strings to dry, and then sold them on the street. But the strings of mush-rooms all looked the same from lady to lady, so maybe they were not homegrown. That could be a blessing, because we read several hundred people die each year from mushroom poisoning after foraging in the woods. In any case, these mushrooms were terrific, very full flavored, and a bit smoky. We would buy some every trip. Then, after several years the ladies were no longer on the street selling mushrooms. We were told the mafia ran them out of town. How much competition could they have been? But the mushrooms sold in the stores were not as good.

Obtaining cash. Renting apartments and shopping locally for food required that we have a good deal of cash with us. Landlords wanted to be paid the first night, and they wanted US dollars. The market mer-chants and the ladies on the street needed rubles. This was before widespread use of credit cards and cash advances. We were warned that US dollars needed to be in pristine condition, both for the landlords and for the bank where we would change them into rubles. Today, you can get cash advances in local currency at ATMs, although you may be charged for the transaction and interest on the money until you pay your monthly bill (unless you use a debit card rather than a credit card). There may also be a limit as to how much cash you can obtain in one transaction or day. But if you are bringing US dollars, best that they be pristine, at least for some countries.

Other Arrangements

We stayed in a variety of facilities in the other countries, and having local colleagues help identify these arrangements was very valuable.

In Latin America, we used hotels usually selected by the foundation's President or local representative. Ultimately, the foundation bought a house in one country which its local representative stayed in, and it became an office and training facility.

In China, we and our colleagues stayed in apartments that were selected by a local acquaintance or in hotels that our American Chinese-born colleague arranged. In Kazakhstan, we used a small boutique hotel.

In all these countries we ate in restaurants, usually with local colleagues who selected them and could guide us through menu items, especially if it was not in English.

The Sky May Be Falling

In one Latin American country, Chris and Bob plus several US staff members stayed in a cute local hotel. It had a large open-air garden with a small fishpond and a covered living and dining area, surrounded on the perimeter by guest rooms. Our group essentially filled the rooms on one side of the garden.

Every evening after work we all would gather in the living area where the hotel provided each of us with a glass of wine served by a nice lady. We tipped her after the first night's "Happy Hour," and the wine flowed quite freely every night after that.

Also, the morning coffee at breakfast was especially good. We were told it came from a local plantation and was delivered to the hotel every few days. We asked if we could buy some, and the front desk clerk said yes. We all bought a bag or two – $3 each. There are some benefits to staying at small local hotels.

One night shortly before Happy Hour we were all in our rooms getting ready for the evening's events. Suddenly, the floor and walls shook. Bob decided this was not normal. The building shook again, and he decided it was an earthquake. He ran out of his room and stood in the open-air garden, figuring there might be less building to fall on him there. One or two others emerged from their rooms and calmly wondered what the commotion was all about, and one other was apparently in the shower and emerged 20 minutes later and matter-of-factly asked what was going on. They all teased Bob for panicking and standing in the middle of the foliage next to the little fishpond in the garden. "I think we are on the Ring of Fire," he explained. "They have earthquakes here, buildings collapse, people get hurt or killed."

"Oh, OK," was the matter-of-fact response. "Is it Happy Hour yet?"

Actually, we have now learned that the recommended course of action is not to go outside unless you can get far away from the building and trees. Debris often falls around the building, and trees topple. It is best to crawl under a table in the room. NOW we know this. Be prepared.

Part II

Developing and Implementing a Project

Perhaps the most unusual and potentially problematic aspects of conducting collaborative projects in other countries is developing and implementing the project. This must be done in a collaborative manner, and many of the unique characteristics of the collaborators and their country come into play. We offer some advice and illustrations of these issues, but the details are likely to be unique to each situation.

8   Project Planning and Funding

General Lessons Learned

As indicated in Chapter 2, projects start in different ways. Sometimes a specific intervention or program is well-known in advance and the task is to secure local colleagues and cooperation to implement it. At other times, a specific project is not fore-ordained and designing the project is very much a collaborative enterprise.

Even when the nature of the project is well known, it helps to involve local colleagues in project planning to the maximum extent possible. This can be started with a "blue-sky" discussion about what people would dream should be done under ideal circumstances. This might include free-flowing thoughts about what is needed and desired, what exists now, what kinds of programs or interventions would satisfy those needs, why people think those activities would actually produce the desired results, and how those activities could be implemented. This helps to get a broad picture of the territory and to get everyone on more-or-less the same page. It also produces a more respectful, collaborative partnership and a better project.

A next step is to engage in a formal "logic model process," which basically addresses the same topics but in a more realistic, systematic, and specific manner. You might lead the discussion and ask the questions, but local colleagues and other stakeholders should provide as much of the information and as many of the answers as possible. Of course, you may have information about possible services, interventions, and research-based principles that they do not have, and that information should be contributed for them to consider.

An important component of this process is a "theory of change." This addresses why you and your colleagues think the activities of the project will produce the desired outcomes. Some very applied professionals may believe that providing one or another service will produce the intended improvements. Perhaps it has always been done that way in this country, and they believe it works. They may be correct, but it helps to consider the evidence and perhaps theory or rationale that support this belief. Of course, you may have evidence and experience that another strategy might be more successful, and that should be brought tactfully into the conversation. Not only does this exercise keep the discussion more realistic, but this information likely will be needed to support grant requests.

Throughout, you want to be as collegial and even deferential in style and tone as possible. You need to avoid giving the impression that you want to drop a project into their country, even if that is close to your goal. There is a delicate balance needed between your local colleagues recognizing that they want and need your expertise on the one hand and their own personal and professional integrity and ability to be contributing members of the team on the other. It may be easy for you to "take over the entire process," which can happen if you have an extensive edge in knowledge and experience. Also, you may expect colleagues at home to be uninhibited in disagreeing with you, but partners in some other countries may behave quite deferentially to you even when they disagree, and this also can be a problem as the project progresses. *Keep in mind that it is their country; their context; and their rules, policies, practices, and likely personnel that will define what can or cannot be done.*

The local circumstances will be especially relevant if the project is quite applied in nature and will operate with people and organizations in the local community. You will simply not likely be able to "replicate" an intervention previously demonstrated to be effective in your country in this new context – there will probably be too many relevant factors that will be different and perhaps some procedures that will be difficult or impossible to implement exactly as before. You will need to identify, often without very good empirical evidence, the essential elements of the intervention that are likely crucial to achieving the desired results. These are what we called "crucial elements" or "non-negotiables," a term possibly too strong and with a dictatorial tone, but it helps for everyone to understand the importance of these components. Try to minimize the number of those elements, but be more flexible on all the other characteristics on the way to molding the project together to fit the local context.

In all likelihood, you will need special funding to implement your project, and it is wise to have at least a general idea of where the money will come from as the project takes shape. You are likely to plan a very different project if you think you can get $200,000 or more than if you think $50,000 is all you can get to fund it. Alternatively, you can plan a "blue sky" project and cut it back when you face financial reality, but it is probably best to acknowledge that strategy at the outset. Also, different funders have different priorities and emphases, and a project may need to be "packaged" one way for one type of funder and another way for a different type. Fairly early in the planning process it helps to explore with different possible funders whether your project fits their priorities or how it could be molded to fit them. Ultimately, you and your colleagues will need to have a fairly specific project plan developed that is targeted at one type of funder before you are ready to write an application.

Funding an international project can be quite difficult. If the project is essentially research, then the traditional sources of research funding may be candidates for support (e.g., in the United States, National Institutes of Health, National Science Foundation; the European Union). These funders sometimes have set-aside funds for international projects, but they tend to support conferences, fellowships, and other kinds of collaborations. If your project does not fit into those categories, then these funders usually "buy" research-based knowledge, and your project must address questions that are of broad scientific interest. Further, you will need to justify why the project should be conducted in a location other than in the funder's country. There are other government funding agencies (e.g., USAID in the United States), international organizations (UNICEF), and foundations that will fund a broader range of projects, and they are more likely to embrace applied projects. Governments and non-profit organizations in the county sometimes fund specific kinds of projects, but the amounts of money are generally much smaller.

Chances are that you will be more experienced obtaining money for projects than your local colleagues, although there are many colleagues in Western Europe, Australia, and Asia who will be quite knowledgeable and experienced in obtaining money from local sources. Consequently, writing grant applications and ethics reviews will likely be primarily your responsibility, although your local colleagues will have to modify and submit them to local authorities.

Creating a budget for an international project can present some unique challenges. Salaries, consulting fees, fringe benefits, inflation rates, hiring practices (nepotism), and other aspects of the budget may be different from what you are used to in your country. A first consideration is the possibility of an immense difference in salary levels between you and your local colleagues, so you may want to make a practice not to share your own budget with your foreign colleagues (unless they are the funder). This practice may complicate the premises of full transparency and equal partnerships, but the differences in what the grant may pay you versus your colleagues can be staggering. It is important to understand these possibilities and accept that there will be this "elephant in the room."

Additionally, if your grant will be paying salaries and expenses in the local country, you will need to determine how the money will be transferred and accounted. This can be challenging in less-affluent countries with less stable banking systems. If you are employed by a major research university or private organization accustomed to international projects, they will have experience in these matters, but be prepared for restrictions and new procedures.

It helps to get a good idea of the roles of each participant on your team. Who is responsible for each aspect of the project? A good team is one in which there is equal responsibility and influence but not necessarily equal work. It should be like a good marriage in which each partner has equal authority and responsibility, but the division of labor may be quite different. This ideal of equality may be very difficult to achieve in practice, because the person who got the grant and must report on its implementation inevitably has a disproportionate amount of responsibility and power. Just try not to overtly exercise that power unless absolutely necessary.

Politics and policies can suddenly occur or change and influence a project. Politicians in some countries may be in power for uncertain lengths of time. They can propose exciting program possibilities, only to be suddenly out of office and influence. Also, political actions by a government can determine whether a project lives or dies, often precipitously. Planning new and grand projects with political administrators can be very exciting but recognize how temporary their position of power may be.

These issues can be a continuing source of friction, even in a well-maintained collaborative enterprise. Although you and/or your colleagues may have considerable experience leading collaborative projects in your respective countries, circumstances can change, and new issues can arise during the course of the project that need to be solved.

Illustrations

As noted in previous chapters, it is important to get to know the country, the organizations with which you will work, and your local colleagues as much as possible before intensive project planning is conducted. In addition, it is very helpful to observe the operations of an agency in which you are likely to work and meet with its staff. So, to the extent possible, try to understand how the agency works now, what it does successfully and what may need improvement, and what kinds of changes might be possible.

Delicate Questions

Suppose you are considering designing an intervention in an agency that provides services for vulnerable children. It may be helpful to know if current practices represent cultural differences or grew out of necessity due to the traditionally low resources available and the limits of time service providers had to care for so many children. For example, suppose you observe during your visits to the agency that there was no talking with the children at the table during mealtimes. You may want to suggest that conversations between the caregivers and children are important opportunities to enhance verbal, cognitive, and social skills in the children and to build caregiver–child relationships. However, you may be concerned that this practice is a difference between cultures rather than an institutional policy or practice. This has to be handled somewhat delicately so you do not seem judgmental, but at the same time, you hope your colleagues know you well enough not to be offended. You might explore this issue in the following way.

"How is your son doing in school nowadays?" you ask one of your colleagues during a lunch break. "Does he talk about school at dinner?"

"Well," your colleague answers, "when asked, my son often says 'nothing' happened at school. He's rather quiet."

Not satisfied with that answer for your purpose, you continue. "When I was growing up, adults discouraged us from talking at the dinner table, especially if guests were there. Is that something parents in your country would do?"

"No, certainly not in my home," a local colleague countered.

If you feel reasonably comfortable in your relationship with your colleagues, you might admit the reason for the question. "I noticed that there are no conversations with any of the children during mealtimes. I take it that this is not a cultural tradition, so can we suggest a change in that during our training of caregivers?"

"We hadn't noticed that before, but it's a good idea. Of course, we should tell the caregivers to talk with the children during meals, because it will improve their relationships with the children, which is more important than rushing through a routine meal."

The point is that all the time spent in preparation and getting to know your colleagues can help you weave through some potentially delicate issues rather quickly. You may need to be careful about minimizing seemingly judgmental comparisons with practices in your own country. What might seem like an insignificant behavior could become the basis for hours of debate if you do not have the mutual respect for each other that permits more frank discussion.

Additionally, it helps for you to get to know the background and training of the team members and their philosophies and beliefs about various relevant issues. Commonly held insights help produce more efficient meetings as well as the logic model discussion. This is critical because the logic model document will become the road map for designing the intervention and its evaluation. You want to eliminate or at least minimize unknowingly offending your colleagues on various aspects of the potential intervention. Agreement on general themes is a good start, because the details of implementing them can be more problematic.

The Logic Model

If you are collaborating to create a new intervention or service, for example, a logic model process should probably be conducted. This typically includes all the major players in the project, and its purpose is to design all the details of the new project – its general goals, its desired outcomes, major elements of the project, and details of its implementation, personnel, and budgets, etc. The logic model process requires a facilitator who understands both sides of an argument, is balanced, and is trusted by all the participants to keep the discussion going and to get to consensus. From the start of this process, it is important to agree on the ultimate goal, and to remind each other of that goal when discussions and decisions get tough. Yet, professionals, even those with similar general values and goals, can have their own ideas about the specifics of implementing an intervention that challenge years of practice. Disagreements are perhaps inevitable.

Consider the following scenario. It was after midnight on the third day of the logic model process. Participants were eager to be finished, and energy was fading; everyone was tired. A few final, but critical, decisions had to be made.

"Let's finish up the last, and in some ways the most sensitive, decisions," Chris, who was selected to be the logic model facilitator, said with as much enthusiasm as she could muster at this hour. "OK, remember we all agreed that every child deserves to be in a family environment, but that is not always possible in every situation. We also agreed that our goal in this project is to create a family-like environment for those children who are served by this agency."

"Yes, and we have discussed what a family is like in terms of size and makeup," steered Bob. "Now, we also have to come to some agreement on how to include children with profound disabilities into the groups."

"Currently they are separated in a group without typically developing kids," observed Chris. "Closely connected to that is the fact that all of our children are in groups with children of similar ages. Is that typical of a family?"

"No, that's not a family!" Susan asserted with authority.

Jim defended the status quo by empathizing with the caregivers. "We will get push back from the workers if we include children with severe disabilities in their groups. They will say those children cry a great deal, they cannot be improved developmentally, and they will disturb the typically developing children who are in their group!"

"Also," added Jane, "caregivers are comfortable with the age group they are assigned to, so those working with only toddlers who are typically developing will refuse to have other children in their group. 'I don't know how to deal with a child with a disability,' they'll argue."

Chris backed up the discussion in the face of this resistance. "Recall the general goal we agreed on – we want a family-like environment. So, what would a family look like? It would not have 12 kids all the same age! No woman gives birth to so many at one time! Few families have all children with disabilities either! We agreed at the beginning to the characteristics of a family; let's put this discussion of what to do in the agency into that context."

"Yes," Susan chimed in. "We agreed that a family may have one to two children with disabilities and certainly would have children of mixed ages, even if there are twins. However, we must consider the feelings of the current providers who have not been trained in caring for children with disabilities or groups of mixed ages, but I agree that these are changes that we need to make to our groups."

After some more discussion, all decided the benefits of the new mixed groups outweighed the concerns, but it would take some effort to implement them. Jane suggested, "OK, let's be sure to talk with the staff before these new policies are announced and assure them that they will be trained and mentored on the skills they need."

The lesson here is to recognize the important and hard work of a thorough and participatory logic model process. No one person should dominate, and all opinions must be considered before trying to come to consensus. If the process ignores serious misgivings about components of the approach, those doubts will come back to haunt the project later, possibly years later.

Further, in the logic model discussion we delved deeply into sensitive discussions of what we as a team wanted to accomplish, what was not working, what needed to change in caregiving practices, what resources existed, and what more was necessary to do. These did not always go smoothly, and our prior insights into and respect for our similarities and differences helped us to be less defensive in disagreements and less dominating in our proposals. We did not take criticisms and disagreements personally. It also facilitated our role in leading the sessions and in making decisions when there were opposing views on some details. It turned out that we all agreed on the important elements, our colleagues knew what they wanted in general terms, and we knew how to design an intervention. This permitted a comfortable comradery to emerge, even in the face of some disagreements, that frequently allowed us to work well into the night sharing our personal and often humorous stories that related to the work at hand but also staying focused on our goal for a strong effective project.

A Funding Epiphany

Funding a foreign project is often difficult, especially if its primary purpose is humanitarian. Imagine the following interaction.

After wine and lunch, Chris and Bob settled into the long airplane ride home following a project-planning visit and logic model sessions.

"Well, you did a terrific job guiding the planning of this project," Bob complimented Chris. "It was a great blend of what they needed, what they wanted, and what we all knew would improve the development of the children. You covered everything, including their budget. But we are missing one crucial detail: Where are we going to get this kind of money?"

"I'm not sure," Chris agreed. "And I only planned THEIR budget. If we add estimates of the cost of our time, other personnel, travel, and all the other expenses we usually put into a budget, we are into 'real money,' like several hundreds of thousands of dollars per year."

"This is a humanitarian project that proposes to improve children's development and lives," observed Bob. "Who funds that kind of thing?"

"In the United States, there are federal, state, and local governments that fund such projects, even a few foundations and combinations of these."

"But they are not going to fund this kind of project in another country," Bob contributed the obvious. "Are there government agencies that fund international programs?"

"Yes. USAID," said Chris. "I saw an announcement of one of their funding programs that would likely embrace this project, but they would contribute only $50,000 per year for a maximum of two years. That will not come close to what we just planned."

"I don't suppose there are any other local sources."

"I doubt the local national or city governments will support a project like this. They do not have the extensive network of human services that we have, and they don't have this kind of money. Also, I doubt they would be interested in a project that involves Americans. They figure Americans have loads of money."

"I just cannot conceive of humanitarian funders coming up with that much money," concluded Bob. "The only organizations I know that give those amounts are the US research funders, such as the National Institutes of Health, the National Science Foundation, and others. But they fund science, not humanitarian projects. Basically they 'buy' research-based knowledge of a general nature."

"Well," paused Chris, "can we make science out of this?"

"Mmmm," mused Bob. "This may take more wine or maybe brandy. They won't fund a project that is 'better care for children in another country.' We've got all kinds of demonstration interventions in the United States that are doing that for American kids. We will have to be very creative to perceive how an applied project could be configured to answer a scientific question. For example, what is scientifically unique about our project?"

"The children are much worse in their physical and mental development than our kids."

"True," agreed Bob, "but that alone is not sufficient. Besides, there are a bunch of reasons why they are so delayed. Their birth status is likely very poor. Their gene pool is probably not the best. Some might be traumatized for various reasons. They might have been abused and neglected. We don't know the nutritional value of the food they are fed. Their medical care is probably OK by local standards but may not be by American standards. The physical status of the agency would not pass US health and care standards, and the list goes on. That is why some people have made the claim that the well-known poor development of such children can be due to any one or all of these circumstances. We can't possibly tease these potential causes apart."

"But don't all these negative factors make the children unique? They are at-risk for nearly every possible reason."

"Yes," concurred Bob. "But we need a basic question to answer. The logic model process we just finished on this trip pointed to a strictly behavioral intervention that focuses on improving caregiver–child interactions that may improve the physical, cognitive, and social-emotional development of these children. That's kind of revolutionary. So maybe the basic question is: Is improving caregiver–child interactions and relationships enough to improve their development in all these areas?"

"And is it enough for these children who may have one or more of all these adverse circumstances in their backgrounds?"

"Exactly. So, the scientific principle we would study is that regardless of their varied adverse backgrounds, these children are massively delayed primarily because of very poor quality or non-existent caregiver–child interactions. We would have to argue that improving those interactions ALONE can improve the physical, cognitive, and social-emotional behavior of essentially all these different children, even those with a variety of disabilities. We don't change nutrition, medical care, safety, etc. – just caregiver–child interactions and relationships. It's a real test of Bowlby's conclusion that such children 'lack mothering' plus contemporary attachment theorists' emphasis on the crucial importance of early warm, sensitive, responsive caregiver–child interactions and relationships."

"Will that fly at the National Institutes of Health?" asked Chris.

"It's worth a try. But if we applied to NIH we would have to add a whole basket of assessments of the caregivers, their behavior with the children, and the physical, cognitive, and social-emotional development of the children. The cost just went up, but NIH can probably afford it if we are conceptually persuasive and have a reasonable, if large, budget."

"But we would have to justify doing this in another country," added Chris. "I suppose we have to argue that there is this population of extraordinarily at-risk children and limited caregiving all in one location. We could not do this project in the United States. We would have to visit each family separately, and the caregiving before and after the intervention would not be similar across children. We would have much better scientific control and for less money doing it in this foreign context."

"You sold ME, but I don't count."

"I think we also may have to argue that such a project would have implications for improving services and policies in the United States," added Chris. "For example, early care and education programs currently emphasize early learning and preparation for school, but this project would highlight the role of early social-emotional experience and relationships with caregivers. It would also demonstrate that one of the reasons foster children do not do so well is that they often move from family to family and do not form relationships with one foster family."

We decided to give it a try. We contacted a program officer at the funding agency and visited her to determine if such a project fit their funding priorities and to obtain any suggestions for how to craft the application. She was very supportive and helpful. We had to return to work with our foreign colleagues to flesh out the project, especially to determine the assessments we would propose, which this particular funder would require. We also contacted some of our American colleagues who were experts in various aspects of this endeavor that were less familiar to us (e.g., the physical growth of such children) and solicited their advice and agreement to be consultants to the project. We then wrote an application.

Not a Good Beginning

Sometimes a project starts quite differently. Consider that we were invited to help a funder evaluate a program the funder had already decided upon but had not fleshed out its details. We had no relationship with any of the players, and yet the first visit was characterized as a working meeting with important decision makers in the country. We needed to describe our previous work and to move into a logic model session to get everyone on board. This high-speed approach made us a bit uncomfortable, because we needed to first visit the targeted agencies and to meet with their directors. We requested these preparatory steps and the request was accommodated, but the tour of the agency was somewhat cursory, and translations were difficult. Suppose this is how the process went.

Upon arriving in the country for the first time, we met with our sponsor and country host to receive our agenda for the trip.

Bob asked, "Is there a way we can visit the agencies before we meet with the key stakeholders, who include politicians and human service directors?"

The local Host replied, "No, they want to know the details of what we are planning before they agree to the work we are proposing in broad terms."

Chris jumped in, "But we need to see the institutions and talk to the directors to know what kind of assessments we could conduct to evaluate the intervention. Also, we should develop the details of the intervention together through the logic model process or there won't be the high level of buy-in we need when times get tough."

The local Host reluctantly acquiesced. "I'll try to set up a quick visit before our meeting with the stakeholders, but it will be difficult."

We eventually visited the agencies before conducting the logic model process. But it did not go smoothly. We had almost no translations, and we were given only one afternoon to do the logic model discussion.

Our evening debriefing after the logic model meeting was troubling. Chris was obviously frustrated. "This is not ideal in nearly any way. What happened in that room? There were so few questions! Are they not interested? Did we walk into a quagmire?"

"Yea, maybe," murmured Bob. "I'm still reeling from the level of hierarchy here, which is not the partnership style we usually seek. We did not have time to get to know the participants in the room, and some of them viewed us as the experts who came to tell them what to do! And at the same time our foundation sponsor, not us or them, is clearly making the decisions. OK, what now?"

Chris volunteered, "We are lucky we have something of a game plan from our previous experience in our back pocket. Tomorrow, we fill in the logic model with what we heard, saw, and know about such projects and present it in hopes of getting feedback, agreement, and buy-in."

We worked well into the night creating a draft logic model for the next day. It was mostly accepted by the stakeholders as a general roadmap, but they never seemed to own the plan or know how to use it, because by and large they really did not plan it collectively with us.

We had to get used to having a different role in this project – despite the fact that we had substantial experience in improving services for children, this was someone else's project, and it was not fully developed yet. Further, our primary role was to design and implement an evaluation.

This experience demonstrated clearly to us that spending "upfront" time in a country with relevant stakeholders, the sponsor, and staff, and getting to know the culture, the history, and the organizations' priorities, were critical to developing a true partnership. The logic model process is a tool that can contribute to building better relationships and a more equal partnership, but this groundwork is necessary to build trust, relationships, and a spirit of working together.

A "Best Obtainable" Evaluation

Evaluations must fit the program they assess, but in the illustration above the problem was that the intervention itself was not planned with the necessary specificity when we started, in part because the logic model process was very brief and not an important step in planning the project. So, we had to assume the intervention's goals were very general and simply consisted of improving children's physical and cognitive/behavioral development. Further, we did not know how much money might be available for the evaluation. In our experience, most people who plan interventions are shocked at the cost of evaluation and have a difficult time allocating much money to evaluation when those funds could provide the intervention to more children.

A hallmark of the evaluation projects that the Office of Child Development had performed for domestic projects was solving this kind of problem. Obviously, the evaluation must ultimately determine if the intervention successfully achieved its goals, but it needs to do that within the very substantial constraints of available funds, personnel, time, limitations dictated by the circumstances of the project, local community characteristics and policies, and scientific credibility. Thus, many compromises must be made on the way from an ideal to a practical evaluation.

The first limitation was the absence of a comparison group of children who do not experience the intervention. We would have to use pre-intervention assessments as the no-intervention comparison in a pre-post-intervention design. The second issue was that children were different ages at any one calendar date, and they came and left the service at different ages and with different lengths of exposure to the intervention. There are some ways of handling this problem, such as using standard scores for develop-mental assessments and some special statistical techniques. Third, assess-ments cost money, especially if each child must be individually tested by a trained assessor. So, the number of different assessments had to be limited, the service's routine measures of physical growth (height, weight, head circumference) had to be used, and independent assessors had to be trained to test children's general behavioral development and assess caregivers' interactions with children.

Fortunately, we found a local faculty colleague who was an experienced researcher who specialized in social psychology. He knew what research and assessments were supposed to be like. He was also a priest, which gave him immediate credibility and respect from the administrators of the agencies, who were often nuns. He identified four students whom we trained to give a standardized infant developmental test of cognitive and behavioral development. We trained institutional staff to conduct physical assessments in a more standardized manner, and we created forms for data to be collected and entered into a database.

Conducting assessments on children was a little easier than assessments on service personnel. Children were present every day, but service personnel often worked only every third or fourth day – they were moving targets. So scheduling assessments was quite difficult. Some degree of preparation of those personnel was needed to ease their anxiety of being observed and "scored" in some way.

Also, there were numerous little factors that had to be accommodated. For example, the hours that personnel worked had to permit them to arrive and depart the agency during daylight hours, because it was too dangerous for them to travel by bus in the dark. Also, personnel and assessors should not be paid on Fridays, we were told, because most people are paid on Fridays, and some get robbed shortly after leaving the bank.

This all illustrates how research ideals must be modified when working in an applied context in another country. The task is to get the best obtain-able, if not the ideal, evaluation results under the practical conditions within which one must work.

I Can't Watch Anymore

Suppose another scenario. We were asked to provide training to the service personnel, especially on how to interact effectively with children with disabilities.

We began this task by observing caregivers in the playroom and in sessions of physical therapy conducted with children with severe disabilities. At times what we saw was difficult to watch. For example, in a room filled with loud music, Chris gasped as the physical therapist, with no warning, vaulted a young child with extreme physical limitations up on a large therapy ball. That was not the worst of it! She then rolled him back and forth quickly with no rhythm, no verbalizations, and no advanced notice. His muscles contracted, and he flailed on the ball in an involuntary reaction to the dangerous speed with which she rolled the ball. His face grimaced in pain and fear, he yelled out but was ignored.

Chris, who had some training in this field, whispered to Bob, "I can't watch any more. It scares me what could happen to that child, and worse yet, the therapist is doing it to impress us!!! What happens when no one is observing?"

The issue is whether we say anything to anybody. Chris felt the danger to children from this practice was substantial, and we were brought here presumably for our expertise in such matters. We decided something had to be said. Because we had no authority in this agency and were viewed as guests, we decided that rather than addressing the therapist directly, we would seek out our Host to relay our concerns. We were hopeful that, because she had a good relationship with the Director of the agency, she would handle it sensitively. The Host reported it, somewhat tactfully to the Director, who defended the technique as standard procedure here. We were then asked to meet with the Director to talk about the issue.

Chris cautioned privately to Bob, "We have to be very careful here. We were simply permitted to observe and yet we submitted a strong criticism of a standard practice. I know that's against our usual operating protocol, but this was an extreme situation in which the child and others could be seriously hurt."

Bob agreed. "Yes, we ought to be calm in our demeanor and deferential when possible. We can't come off as the big shot experts from the United States who came in to blast them."

"Agreed. Maybe instead of focusing on what they were doing, the most tactful and diplomatic approach would be to communicate that therapy for children with severe disabilities is done quite differently in the United States, and we thought they might want to know what we would do with such a child. In the process, we can mention why we think what we do is best for the children. If they conclude that they should consider alternative methods, we can then offer to provide a workshop on our techniques that even staff untrained in therapy could use."

This somewhat tactful informative style was used during the meeting with the Director and supervisory staff. All partners left feeling that the training should be offered to improve the situation, not to be disparaging, and that there would be much useful advice offered. There was agreement that all staff, not just the physical therapists, would be required to attend this "hands-on" training workshop on how to feed, position, and transport children with severe physical disabilities. Thus, therapists would not be singled out, and perhaps the children would receive improved care. The workshop went very well, and participants felt the techniques offered were useful.

What we learned from this situation was that because we had no relationship with the indigenous staff and were in a "visitor" role, we typically would use our observations to design the intervention. But we were not designing the intervention. Nevertheless, because we deemed this case to be unusual and extreme and feared for the safety of the children, we felt some responsibility to do something. Using the relationship between the Host and Director was a circuitous but effective way to accomplish what was needed. The consequences of going directly to the staff person, Supervisor, or Director could have resulted in being viewed as interfering, obnoxious, demeaning, or worse, and we may have even been asked to leave the premises with no hope of working together in the future and thus no opportunity to improve the situation for the children involved. Further, emphasizing what is done in the United States as an informational approach rather than directly instructing the local therapist avoided conflict and preserved relationships.

Guest Workers

Consider a different situation. While the intervention was still under development when we arrived on the scene in the previous illustration, imagine that a foundation had already developed an intervention that was ongoing in this next example, so it was clear from the beginning what our role was and precisely what intervention was to be evaluated.

Initially, we were able to establish a relationship with a faculty member at a local university, who arranged visits to several services and other facilities to give us a general impression of care for vulnerable children in the region. Of course, these facilities were not a comprehensive sample of services, but they represented a variety of different approaches to caring for children with different needs. Caregiving in many of the services was similar to what we had observed in other countries; namely, children were cared for in large homogeneous groups with too few and changing caregivers who performed the necessary caretaking duties in a business-like fashion with little social or emotional warmth. Even the food served the children appeared to us to be very basic – mostly thin soup.

Medical care also appeared quite "traditional," that is, sometimes using techniques abandoned long ago in the United States. For example, we visited a hospital clinic for children with a variety of medical problems, including those with different kinds of disabilities. The treatment of children with cerebral palsy and Down Syndrome, for example, emphasized physical therapy consisting of forced movements of the affected arms and legs, sometimes producing screams and crying in the children. It also included bracing of limbs to provide more mobility. Some of the practices we observed may provide some benefit if carried out consistently. But there was little chance that any of the biweekly or monthly techniques used at the clinic could be used consistently between visits, because the parents and other caregivers were not permitted in the room where the therapy was provided. We also knew of US research that showed that even more modern physical therapy was less effective than early intervention techniques for these children.

More specific to our intended purpose, we visited a facility to observe the foundation's intervention in action. The quality of the caregiving provided by the agency's staff was similar to the other agencies we had visited. There was little caregiver–child social or personal interaction or play. However, there was one difference: There seemed to be quite a few more caregivers available than we had seen before. We later found out that several caregivers were "guest caregivers," that is, women from the community who were "on call" to supplement the regular staff when the agency had visitors. This practice gave the appearance of smaller children:caregiver ratios and thus better care than was typically the case. It was a warning sign that hosts and agencies can put on an artificially good face for visitors.

Around mid-morning the foundation's special caregivers arrived, perhaps six to ten of them, each wearing a brightly colored tee-shirt. Each sat on a small rug on the floor with a few toys, and they spaced themselves around a large room. Then, the regular staff brought in the children, and each special caregiver received three to four children who were regularly assigned to her. These caregivers played with their small group of children for perhaps 60–90 minutes. They did this on several days of the week. The play could be soft and slow or fast and intense, depending on the particular caregiver and her children. The regular caregivers were not involved in this activity, nor were they trained to do it when the special caregivers were not present. We thought training the regular caregivers by example to do the same kinds of interactions might provide a more consistent and permanent strategy, but it was not our role to suggest changes in the intervention program. In any case, the special activity constituted much more interaction with a consistent adult than the children would ordinarily experience.

We visited other agencies and observed the sponsor's intervention implemented in more-or-less the same manner. In one, our colleague was able to roam from room to room. He went into one room, which was not shown to us, that consisted of children of mixed ages who were all in their beds. They either had various disabilities or looked quite sick or even comatose. There was one caregiver for all the children in the room, and she did not interact with the children except as necessary.

Our colleague concluded that staff in these services had the same beliefs as existed in several other countries, that is, that there was nothing that could be done for children with disabilities. Children with such severe limitations could not be improved and were not expected to live very long. The agencies were short-staffed, and perhaps the administrators assigned their limited staff to serve children who could be improved. We had experienced that same attitude about children with disabilities in other countries. Such children were also housed together in the same room with essentially no expectation that they could be improved in any way, and we also were told that some of these children died while in the service.

Our evaluation plan had fewer obstacles than the previous illustration. For example, to be able to have a no-treatment comparison group for our evaluation, we asked if there was an agency in which the foundation's program was planned but not yet implemented. This would at least permit a before and after comparison within the same agency. Yes, there were two such agencies in a somewhat remote city. It was largely a manufacturing city, and the air pollution was immediately noticeable, almost suffocating, getting out of the car and walking into the hotel. These two agencies had already agreed to have the foundation conduct their program, so there was no need to build interest and persuade staff to implement the intervention. Besides, regular staff did not have to implement this intervention. Instead, they got a break from caregiving while the special caregivers played with the children.

"Collective Foster Care"

Sometimes the intervention needs to fit a variety of somewhat unique local circumstances. China, for example, had begun to explore foster care for several years in collaboration with a British non-profit organization and other groups. In contrast to many other countries that have difficulty finding adult couples to foster children, many Chinese couples eagerly wanted to raise another child after the one-child policy restricted them to a single child. Although placing children into families in the community seemed to work well enough, it challenged the Chinese social-political desire for central control, because monitoring individual families spread throughout the city was physically difficult. So, a sort of "collective fostering program" (our descriptor, not the Chinese's) was being explored.

We saw two such attempts. In one case, very basic communal apartments were established. An "apartment" consisted of two adjacent rooms in which an adult couple would live and care for three or four young children. These "apartments" were not private, and there were several, one next to the other, on a single floor of a building on the agency's campus. Each foster family was supported by all the facilities of the agency, and this arrangement permitted monitoring of these families by the agency's professional staff.

We also visited another agency elsewhere in China that had a very large campus. They had the space and resources to build a building with 24 self-contained, private apartments, each consisting of several rooms. Each apartment was given to an adult couple who cared for four foster children.

The man could hold a daytime job, the woman did grocery shopping and domestic chores, and the family received additional supports from the agency. This arrangement was very attractive to potential foster couples who got a modern, new apartment to live in, and it permitted the agency to supervise as needed.

After a tour of the agency's facilities, we met with the Director and Managing Director. They expressed some interest in demonstrating that their new foster family program was improving the children's development. We were also very eager to explore this possibility, because at the time, to the best of our knowledge, there were no empirical demonstrations of the effectiveness of foster care to improve children's development in countries that traditionally relied on institutionalization.

But we did not have much, if any, money to support such an evaluation. Could an evaluation be done "on the cheap?" We knew from our previous work and research by others that a social-emotional intervention such as foster care should improve young children's physical growth, including height and weight, in addition to their cognitive and behavioral development. Did the agency routinely measure the physical growth (i.e., height, weight, head circumference) of children in the agency's traditional program as well as those children who transitioned to this special foster care program? Yes, and they had done this for years, permitting growth comparisons of children who remained in the traditional program with those who transitioned to foster care. In addition, we arranged to send our Chinese assessors to conduct ratings of the sensitivity and responsiveness of the foster mothers' interactions with their children to correlate with children's growth improvements, providing some justification that it was caregiver–child interactions that were related to the children's growth improvement. Further, the Director volunteered to house and feed our assessors during their visit to the campus to make these assessments and gather physical growth data.

Here was an example of how a project developed unexpectedly, from beginning to end, with minimal financing and good local collaboration.

Promising Opportunities, But...

This example illustrates how exciting and disappointing plans floated by a policymaker can be.

We were fortunate to meet with several national government administrators in one country to explain our background promoting the development of vulnerable children.

One such meeting was with the minister who was in charge of all child welfare in the country. He was partly educated in the United States, and he was very forward looking and creative in his policy perspectives and ideas for his country's future child welfare system. He seemed impressed with the unique purpose and structure of the Office of Child

Development at the University of Pittsburgh, our contacts with major US academic and professional specialists in certain aspects of child welfare, and apparently us personally. He suggested that we consider proposing a national collaboration that could be located at the country's major university that would bring US experts in child welfare there for short periods of time to lecture and consult with local academics and government officials. He thought he could fund it.

We were very excited about this possibility – but only for a few months when we were told he was no longer his country's minister in charge of child welfare. Government officials may have uncertain tenures in office.

This kind of event can happen in many different countries. In another country, we met with the country's minister in charge of education. He also had been partly educated in the United States and had innovative ideas for the country's educational systems. He listened attentively to our very brief description of our previous work and our plans to provide training to the local professionals caring for vulnerable children and adoptive parents.

"This should not stop with this," he observed. "How can this be expanded?"

"It would be best," Chris ventured, "if we could train a few faculty members at your universities to provide courses in the subjects of the training that we provide to this small group of professionals. We and some of our colleagues from other US universities could provide reading material and syllabi and even have your faculty visit the United States to see how these techniques work in US service agencies. That way you would have a continuing group of trained professionals to work in a variety of your services."

"That's a great idea. I'd be willing to fund that effort."

We fleshed out this idea informally over the next several months, only to learn that this minister somehow fell out of favor and no longer had the responsibility or the authority to pursue this idea. Once again, politicians' power is sometimes ephemeral.

An Emerging Plan

Sometimes projects are not planned in advance but rather emerge in conversations with local funders and professionals. Further, they may be funded by a local organization or individuals who are also still identifying their priorities and not experienced in traditional grantor–grantee processes.

Consider the following scenario. We made a trip to another country to visit with a small group of wealthy citizens who had recently created a foundation and were eager to improve the development of vulnerable children in their country. We were to visit several services, some of which were funded by this group, and work with them to develop the next steps in their planning. Presumably a collaboration might emerge that they would support.

First, we visited two institutions for infants and young children in two different cities. The institution in the capital city was in a large, well-appointed building with relatively spacious rooms and reasonable physical facilities. We learned about the number of children and caregivers, staffing schedules, and other operational characteristics. These data plus the nature of the caregiving were similar to what we had seen in other countries.

The other institution was located in a small city that was a five-hour drive away. We made the trip in the dead of winter through virtually barren terrain ("tundra"?) on a bumpy road being repaired. We saw no living thing the entire way, except a few horses and scruff peeking through the ice and snow. Then, suddenly we entered this small town in the middle of nowhere, or so it seemed to us. The institution was in an older building but physically adequate, and the particulars of the children and caregiving staff were roughly similar to those in the capital city.

The institutions were administered by the government, most immediately by the city health ministry. Consequently, the foundation did not have any official influence over them. The government perceived the institutions as providing good care, despite strong research evidence to the contrary, which they may or may not have known about. The foundation hoped we might convince them that the care in the institutions could be improved and children would benefit.

Back in the capital city, we also visited a type of half-way house for children and youth who might have been abused or neglected, found on the street, gotten into trouble with the police, and so forth. They would be brought to this agency for temporary housing while their situation could be investigated, mainly to determine if their parents could be found and were able to rear the child. The building was quite basic, and there were only a few staff either to care for the children who were temporarily housed there or to investigate their backgrounds and parents. Those children who the authorities judged could not return to their homes were sent to the Children's Home.

The Children's Home had a large campus and housed children approximately 4–18 years of age either in a large building or in groups of 20 in three-story "fraternity houses." These houses consisted of living, dining, and exercise rooms on the first floor, and bedrooms, baths, toilets, and all-purpose rooms on the second and third floors. In addition, the Home had a gymnasium, a performing arts facility, arts and crafts rooms, and other amenities. Both boys and girls lived together in these houses; they had one major live-in caregiver, a day-time assistant, and staff to transport children to medical treatments and special off-campus appointments. All children attended nearby local schools off-campus.

The Children's Home also housed a set of professionals who provided services for the children and supported parents who wanted to adopt them. The foundation provided financial assistance for these efforts. These professionals, who had training in psychology, social work, and related disciplines, provided us with information about the children's most frequent behavioral problems, how well the children managed at school, and their own needs for additional training on various topics.

We then visited the Mothers' Home, a kind of half-way house for pregnant and new mothers with their infants. The aim was to support the mothers-to-be in a country in which unwed pregnant girls were ostracized. The Home also helped new mothers care properly for their infants and prepared them vocationally and psychologically to keep their babies rather than relinquish them to the institutions. This facility was supported by the foundation.

Finally, we visited a clinic for children with disabilities who were being reared by their parents, which was supported financially by one member of the foundation. Again, the treatment of children with cerebral palsy consisted of a type of physical therapy in which paralyzed arms, legs, and other body parts were forcibly exercised, often with great discomfort and pain to the child. Children were rolled on large therapy balls with muscle extensions that typically should not be encouraged in children with these disabilities. Others were forced to use their arms to hang on bars whether or not they had the strength or mental capacity to do so. Once again, Chris, a specialist in children with disabilities and early intervention, could barely watch. With tears in her eyes, she walked quietly over to another visitor, an attorney from the United States, who looked at Chris and declared, "I can't watch any more, and I don't even know whether what they are doing is right or wrong!"

In a debriefing meeting with the clinic staff and the supporting foundation member, the most that Chris could say was that the treatment approach commonly used today in the United States was "quite different." The Director told us the methods used here were taught to them from another country, but she was willing to invite US physical therapists to introduce other methods. When Chris and Bob returned to the United States they met with some of the top physical therapists in the country, and collaboratively they wrote and presented a proposal to the foundation. No response ever came forth. Once again, disappointment in international work was experienced.

After all the visitations, discussion turned to what the foundation thought they needed and what we might be able to provide. We discussed a variety of needs and how they might be met, and agreed, at least tentatively, on a plan for us to provide training and support to the professionals at several of the facilities we had visited. However, we were the first foreign grant the foundation awarded, so they were new to this enterprise. They were still in the formative stages of deciding precisely what the foundation wanted to do, and other players (e.g., the government, institution directors) often presented challenges to their ideas and implementation plans.

After extensive discussions with the foundation leaders, a project was agreed upon. But the granting process was very different from getting a grant from a large, national funder in which the grant/contract procedures are well established and nothing major changes once the grant is awarded. Instead, the amount of money had to be negotiated, the foundation reduced the number of years from what had been proposed but expected the same outcomes, several technical issues to the contract had to be negotiated with our university, and the project's fundamental program priorities changed over the course of the project (but no corresponding change in the budget). Recognize that when dealing with smaller, private, and local funders and foundations, the process and procedures can be quite different than working with a major funding organization with many years of experience funding all kinds of projects.

9   Administrative Issues

General Lessons Learned

Administering an international collaborative project and grant raises a variety of new issues, even if you have been the director of a unit and/or the principal investigator on several domestic projects. These unique issues pertain to certain sections of the grant request, budgeting and managing expenses to be paid in the other country, contracts with foreign organizations, handling the transfer of funds to international organizations, ethical reviews, and monitoring and supervising the project's progress.

When we started, we were totally uninformed about all these matters despite having a good deal of experience in administering large-budget, collaborative, multi-sited research and service demonstration domestic projects and program evaluations. Our university had some, but not a great deal, of experience in these international matters at the time we began, and we were so naïve that we did not inquire of the university regarding these issues until we faced certain problems. We learned by doing and sometimes stumbling into otherwise preventable quagmires.

Use your university's or employer's resources. Now our university, and we suspect most other major research institutions, has several offices to provide advice regarding all these issues. Our advice is clear: *Contact these offices as early in the planning phase as possible to learn what to expect; the laws, rules, and policies under which you must operate; and how to financially and administratively implement your project.* This should include a preliminary discussion with your own ethics review board, the travel office, the international specialists within the research management and accounting office(s), the legal department, and any professional colleagues who have worked in the country you are considering. There may be a central office devoted to international activities that might be your first stop.

There are numerous laws and regulations pertaining to conducting international projects, even pertaining to data sharing, travel, and other aspects that most of us might not anticipate. The rules multiply if your project is likely to be funded by a government agency. Be aware that some other countries have regulations regarding what data can be collected and taken out of the country (e.g., European Union) that can seriously limit what you will be able to do in your project. It may take more than one consultation with these resources, one at the beginning and one after you are much further along in planning. These consultations will save torment later. You will likely not anticipate all the issues that may arise, but you will know who to consult when they occur.

Plan for what you will leave when the project is finished. If your project is an intervention, plan so it can be maintained locally without additional funds when the project is over. This can be very challenging in cases in which the project budget pays for materials (e.g., special food and nutrition, medicines) that are otherwise not routinely available in the project's location. But in other cases, for example, have the project build local experts who can continue the work when the project ends, and try to plan the intervention in a way that it can be maintained financially on the existing budgets of local organizations. Ideally, you want the intervention, if successful, to remain as a legacy of the project. Otherwise, you look very mercantile – you come in, get what you want (i.e., the project data), and leave nothing in return.

Expect unusual practices from small and foreign funders. Administrative and financial policies and procedures are well established for most large and experienced funding agencies, especially most national government agencies (e.g., National Institutes of Health, National Science Foundation, European Union) and major foundations. But new and small foundations or other funding organizations, especially those residing in some other countries, may not be experienced in funding such projects and may not operate in the same consistent, dependable manner. For example, payments may be late, budgets may be changed, and even the major purposes of the project altered by the funder in the middle of the grant period. Simply understand that such changes can and do occur, and there may be nothing you can do except to go with the flow and try to accommodate.

Be prepared for political influences. Another issue unique to some international collaborations relates to the possibility of greater local political influence. American universities operate under a very broad freedom of information policy. But in many countries the government exercises a great deal more control over what can be done and what information may be communicated, including research results that the government does not view favorably. Small foundations may also limit what you can do and publish (see Chapter 14). While American universities insist on the freedom to publish and communicate, they need to accommodate to the realities of working in some countries. Further, they should not be surprised at government influence over results – it happens in the United States to some extent, for example, when government agencies submit reports that are then reviewed and sometimes "edited" by the Executive Branch to conform to political principles. You may have to exercise some judgment to balance such editing with truthful and accurate representation of your work.

Illustrations

Targeting a Funder

As indicated above, funding an international project can be quite difficult, especially if you need a fairly substantial amount of money. So, the first task is to locate potential funding organizations that are interested in the issues you want to pursue.

As described previously, it helps to talk with potential funders about your possible project before you submit an application. Not all funders welcome such an overture, perhaps feeling it provides you with an unfair advantage over other applicants. But others welcome the opportunity, although there may be limitations on what kind of information they can communicate to you. It helps to read the funder's priorities ahead of time to determine if your project fits them or to perceive how to cast what you want to do to fit those priorities. Be aware that funding priorities change, and a website may not always be up to date. Sometimes we ask a funder to confirm their priorities at the beginning of the meeting to be sure we are on the same page. Your project and a funder's priorities may not match perfectly, but you may be creative in casting your objectives in language that comes closer to that of the funder priorities.

In one instance, after investigating a funder's priorities online, we sent a letter briefly describing our intended project and asking for a face-to-face meeting. The funder confirmed that they were a potentially appropriate source of support for our project and that they would flag the application when it came in. They wanted to know why the project had to be done in a foreign country and then urged us to write the justification in the application. We discussed budget limitations and how to handle them, and they urged us to contact the international section of the federal ethics office for advice on how to proceed with US and international ethics reviews (although your own ethics office should be up on these practices).

A Suitcase of Dollars?

Early in the development of a major project to be conducted in Russia it became clear that we would need to hire several of our international colleagues to be our partners to conduct the day-to-day implementation of it. In addition, we would need to pay numerous assessors, pediatricians, caregivers, and others to perform various functions. This was going to amount to $100,000–$200,000 per year. How were we going to get all that money to them?

We naively had images of carrying a suitcase of US dollars on each trip to Russia and quickly decided that was impractical, risky, and absurd. International wire transfers were possible, but who would we wire the money to? The grant was not going to subcontract with a local university, because the project only involved one faculty member who could best be hired as a consultant. The agency in which the project would be conducted was not really an independent organization, and sending it money would risk getting local government involved. We were told that the Russian economy was in shambles and the banks were unstable. It seemed we were out of viable alternatives.

Fortunately, we learned of the Civilian Research and Development Foundation (CRDF). It was an independent organization established jointly by the US and Russian governments for precisely this purpose. After the fall of the Soviet Union, the US government became worried that Russia's premier nuclear and other scientists who upheld the Soviet military establishment would defect with their knowledge to help other countries develop advanced military capabilities, including nuclear weapons. These countries might be less responsible in handling these weapons than the Soviet Union. The United States thought it was better to try to keep these scientists in Russia by providing them with research grants and contracts.

The US government faced the same issue we did, namely how to handle the money. So, they collaborated with the Russian government to establish CRDF to transfer the funds to buy equipment, pay scientists, and account for the money according to reasonable fiscal practices. The Russian government agreed not to tax the salaries. CRDF also became facile at obtaining necessary approvals and clearances by Russian government ministries for such projects, and they learned their way around Russian policies and practices.

We called the Director of CRDF. Would they handle such a project? We were not hiring nuclear physicists, nor were we going to buy a reactor or some other big piece of technical equipment. The Director listened and asked several questions – it was clear he knew what he was doing and that we could be confident in CRDF's handling of this project. He admitted this project was different than most of those CRDF managed, but it was different in a way that appealed to him. His mother had worked with agencies that served children in Moscow, so he had a special personal interest in this project.

The arrangement was that our Russian colleagues would set up accounts at a specified local bank for each person who was to be paid by the project. They would then send to us a monthly invoice of who needed to be paid and how much, we would approve it, send it to CRDF, which would then transfer those sums to those accounts in the specified bank. CRDF would notify us and our Russian colleagues that the money for salaries, reimbursements, and supplies had been deposited in their accounts, and then the recipients would have three business days in which CRDF would guarantee the money would be available at that bank. It may be available longer, but CRDF would not guarantee it. Why? Banks in the country were folding right and left at the time and could not be trusted to remain solvent! All this provided an accounting procedure that was acceptable to our university and the US government. They also arranged for the project to be cleared by the federal Ministry of Education. CRDF charged us 8%; it was well worth it.

CRDF was a blessing. We are not sure we would have done the project without them under these conditions. But CRDF was unique to US–Russian research projects and was not available for projects in other countries. However, the situation is better now for wiring money to institutions and banks in many countries. But our advice is that before you get too far down the road, especially if it is a big project and you are going to employ or pay local people, work with your own university or employer to determine an acceptable way to transfer and account for the money going to the other country, especially if that country and the institutions with which you will work are less experienced in such matters.

The Hidden Perils of a Subcontract

If you are the grant recipient and you want to hire local colleagues and conduct other financial transactions locally, your university or employer will want a subcontract drawn up with a local institution with which you will work. For one of our projects, our university issued a stock subcontract. It covered the usual points, written in typical administrative and legal language. We showed it to our international colleagues for potential approval.

They had never seen a document like this. For one thing, they did not understand what such a document should cover, and they had difficulty reading the contract language. They wanted a more political and public relations document that would state why this project should be conducted, what good it will produce, and why the local agency and other groups should endorse and support the project.

We tried to explain that in the United States such a contract did not usually cover those topics, which presumably had already been thoroughly discussed and agreed to. Instead, its purpose was to clearly describe who will do what, when, for how much money, and what will happen if those responsibilities are not met. We also agreed that the language was a little turgid, even for native English speakers.

We returned home and explained to the head of our university's Office of Research what we had encountered and asked if a simplified statement that focused only on the essentials written in "plain English" would be sufficient. He took the challenge and produced a short, to the point, readable statement. This document seemed to suffice, and our international colleagues agreed to it.

However, be aware that your international colleagues may not have much experience with US subcontracts and certainly with the language in which they are typically written. Explain what they are intended to cover and not cover, which is only part of the total arrangement and their role in it. Be aware, however, that such contracts essentially define an inequality in your collaborative relationship with your international colleagues. They basically say, "you are working for us, here is what you will do, here is what we will pay you, and here is what will happen if you don't do these things."

Adventures in Budgeting and Hiring

The economic system may be much different in the country in which you will work rather than in your home country, and you will need to adjust budgeting policies to fit it. For example, a usual policy for federal grants to a university in the United States is that they may pay a person's salary but at the same rate of pay as that person would receive from that institution without the grant. Further, that person does not receive any extra money, only the source of their salary changes from the university, for example, to the grant. This policy applied to Chris and Bob; that is, a portion of our salary was paid by the grant instead of the university, and we did not receive a dollar extra as a result of obtaining these grants.

International salaries may be much different than US salaries, and this policy would not work reasonably there. In this instance, the grant was not given to the international university, so we hired our colleagues as consultants and paid them an amount that we agreed on. It was substantially more than their university "salaries," but much less than a US consultant would cost. Our colleagues who managed the subcontract with our approval hired their wives, which we would not have been able to do in the United States, and they hired several of their students in various capacities but not through the university, a practice that was common there. These hires were essentially similar to our hiring of graduate students, but they were not hired through their university.

There also seemed to be a practice to pay people more if they were asked to do anything different from what they were currently doing, even if the new task fell within the general responsibilities of their current job. For example, we needed caregivers to accompany children to their developmental assessments. This would be done during their regular working hours, it did not require any new skills, and to us it seemed to be part of their job of caring for the children. But our international colleagues insisted that the caregivers needed to be paid extra for this task.

We also encountered another unexpected situation when the government raised the salaries of caregivers. This move called attention to the fact that caregivers who were paid out of the grant did not receive a contribution toward their retirement. So, we raised the salaries of caregivers paid out of the grant to match the raises and then added more to compensate for the lack of a retirement contribution.

Similarly, inflation in the project country was quite high, between 10% and 15% per year. We had only budgeted inflationary raises at about 3%–4%, which was common in the United States at the time. We and our international colleagues had to take money from something to provide appropriate raises in the context of this rate of inflation. Study the economic history of the country in which you will work before budgeting annual raises.

By and large, our international colleagues were quite good at budgeting, monitoring expenses, accounting, and conforming to US grant policies, all of which were relatively new to them, especially on this scale. As much as we tried to make this project as collaborative as possible, ultimately the power was not equal between us, because it was a US government grant to a US-based university, and we were ultimately responsible for following all the associated financial regulations and the scientific quality of the project.

Our colleagues understood this for the most part, and we tried not to assert our responsibility/power in this regard. Usually, we simply explained the funder required a particular procedure or we would have a difficult time justifying a specific research strategy to the funder or to journal editors. For example, our colleagues wanted to include several assessments that lacked conventional validity information. We said those assessments would not likely get approved by journal reviewers. This seemed sufficient, and the matter was dropped.

"You Can't Do That. It's Illegal"

As indicated in Chapter 7, we stayed in private apartments in some countries. Most landlords wanted to be paid in US dollars, and we could always expect them to knock on the door the first night to be paid. We were exhausted from the trip and unpacking, and sometimes had to force ourselves to stay awake and wait for the landlord to arrive. They would accept $100 bills, which could be transported easily, but US banks sometimes had difficulty getting unmarked and unfolded bills for us.

Other landlords wanted to be paid at the time of making the reservation, so the agency with which we worked paid this on our behalf. The agency also paid for our police registration and provided lunch for us during our daily meetings. We reimbursed the Director for these expenses, and she gave us an itemized receipt and signed it.

This arrangement was satisfactory for the University Accounting Office to reimburse us for these expenses. Several years into the project, however, we had a meeting with the University Legal Department about another matter related to the project. At the end of the meeting, the lawyers asked a few questions, seemingly out of the blue and unrelated to the topic of the meeting.

"Where do you stay when you visit your project?"

"We rent apartments. We cannot afford hotel rooms, and apartments allow us to cook and work together at night and on the weekend."

"Who do you rent these apartments from?"

"Local citizens who rent out their apartments."

"Are these apartments insured against fire, theft, etc.?"

"We doubt it. The government used to own them, but now the apartments but not the building have been given to the tenants."

"So, if something happened, you would essentially have no recourse."

"Probably not."

"This is not good. How do you pay for these apartments?" the lawyer continued.

"Either in cash to the landlord, or the agency rents it for us and we reimburse the Director of the agency for the rent and other expenses they advance on our behalf. The Director provides us with an itemized receipt."

"You pay the Director of the agency? Who does the Director work for – who pays the Director's salary?"

"The government, we assume."

"That's what I thought. *You can't do that. It's illegal.* A US citizen cannot pay a foreign government employee. Renting these apartments is dicey to begin with. You'll have to rent hotel rooms or find another way to pay for the apartment."

"OK, we do not want to break the law, but renting these apartments has worked very well in the past. Research Accounting accepts the receipts for reimbursement, and they are much cheaper, more comfortable, and convenient than a hotel. We could not afford to manage this project if we had to rent hotel rooms."

"But paying the Director is illegal."

"We are not paying the Director. We are reimbursing the Director for expenses she advanced on our behalf. Instead of disallowing this procedure, is there a way to make it work legally?"

"Mmmm," murmured another attorney in the room. "Could you put a statement at the bottom of the receipt under the Director's signature to the effect of 'I verify that I have received no money personally from this transaction.'"

"Sure, why not?"

"That should handle it."

We have learned from this that we should clear such procedures with the appropriate offices before the project starts. Further, when attorneys or other administrative personnel tell us we cannot do something, we back up a bit, tell them generally what we want to do and why, and then ask how can we accomplish this in appropriate ways and what are our choices? Often, they can readily offer acceptable suggestions, as ultimately happened in this case.

"That's Not Legal in Pennsylvania"

We urge you to contact your Institutional Review Board very early in developing an international project. They are now much more knowledgeable about international projects than they were when we started our international work. We contacted the international wing of the Federal Ethics Office. They were very understanding and helpful. They indicated that we would need to pass the typical review at our own university as well as some review in the location of the project. A university was more likely to have established a process than other organizations, but they recognized that universities in many countries were only recently exploring ethics reviews and may not have developed a committee and process yet. In that case, see if they would set up an ad hoc committee to review the project. The latter was the case, so we asked the Dean of the School of Psychology to have the project reviewed by three individuals.

The Federal Office also warned us about approaching our own Internal Review Board with a foreign project. They suggested that foreign projects were unusual for such boards, and they were likely to approach a variety of issues from the perspective of the United States and not from the perspective of the country in which the project was being conducted. We would need to contain our frustration, they advised, and simply explain the circumstances and how things work there.

After submitting our application to our university's Institutional Review Board (IRB), we were asked to appear in person in front of the entire Board. Basically, they seemed interested in the project, but they were concerned about the usual issues that IRBs focus on.

"Who gives permission for the children to participate?" they asked.

"The Director of the agency is given total authority over the children, every aspect of their health, education and welfare," we responded.

"But that would not be legal in Pennsylvania," they responded. "A foster mother, for example, cannot give permission in our state."

"We recognize the Pennsylvania law, but the Agency Director has total legal responsibility for the children. She can authorize surgeries, for example."

"What about the caregivers? You will observe them and give them questionnaires, but you do not require them to give signed consent. Why not?"

"The intervention and assessments will be conducted on everyone in the agency. They will be part of the way the agency conducts its business. We liken the agency to a school under the IRB instructions which say that informed consent is not required under these circumstances."

"Well, sort of. Could you give all the caregivers a one-page notice that describes the intervention and the assessments, so they are at least informed?"

"Sure."

"What have you done about insuring cultural appropriateness and sensitivity?"

"We are working closely with three local colleagues. They review all the training modules and train the professionals in the agency first, and then they all suggest any changes in those materials before the caregivers are trained. Our colleagues review and translate the assessments to insure appropriateness."

"Are there any minority children and caregivers in this study?"

"We don't control which children or caregivers are or are not included. All children who meet the criteria for being sent to the agency will be included, and that group will include whatever minority groups might be represented in that population."

"One last question, but it's an important one. What will be left when this project is finished? Historically, there have been some infamous cases in which an American project was implemented and produced much better health outcomes in the local community, but then the project ended and the Americans went home, and the country was left without the materials that produced the benefits. They rightly felt used."

"We are aware of this potential problem and have taken steps to avoid such an outcome. First, in planning the intervention we have relied on a train-the-trainer strategy in which we train the professionals using a written curriculum, and then they in turn train the caregivers. This means they are prepared to train new caregivers at this agency as well as others throughout the country more-or-less indefinitely.

"Second, we have planned the intervention so that it can be maintained on the agency's government budget when the project is finished. Our colleagues there knew we had the money to hire additional caregivers, but we insisted that the intervention could not add more caregivers or otherwise cost more to operate than the agency's current budget allows. Most of the money will go to start-up costs and measurements, not operating expenses."

"That's wise and good. Thank you."

Ethics reviews are not the only permissions one might need to operate a project. In our case, we sought the blessing of the Dean of the School of Psychology at a major university in our international location. One of his faculty members would be involved with the project, and we would need him to arrange a local ethics review. Several of us met in his office at 9 am. The meeting dragged a bit, and the Dean did not seem too interested. Then, suddenly a bottle of vodka appeared on the table. The Dean's face brightened. He smiled, quickly got glasses for all, and toasted the project. When in Rome....

There also were administrators who oversaw the activities and operation of the agency who needed to be on board. Our international colleagues briefed city health department officials, who were responsible for the governance of all the child-serving agencies in the city, and they approved it. In addition, two other administrators were contacted, one who was responsible for the medical care and the other who oversaw the education of the children. We visited these two administrators, explained the project, and sought their approval. They listened, asked a few questions, and said, "Fine, go ahead."

Frankly, these two administrators did not seem too interested, or they did not think the project would change very much or improve the children's development. Behavioral pediatrics was essentially unknown there at the time, so a behavioral intervention was not expected to improve children's development much, especially their physical growth. Of course, the point of the project was just the opposite, namely to show that a behavioral intervention indeed could influence children's physical as well as cognitive and social-emotional development.

Working with New Foundations

In some of our projects, the President of a new foundation dictated the project and our involvement in it. After getting a sense of the interventions the funder would implement, we accompanied the Foundation Director to meetings with the wife of the country's President, who was in charge of all children's services, government administrators overseeing the institutions, and the directors and senior staff of the agencies. Our role was to substantiate, based on our previous experience, that such changes would improve children's development. Also, having an independent university-based group that would conduct an evaluation to document such improvement in the new project lent some measure of credibility to the enterprise.

Once an intervention was planned, we submitted a grant request to the foundation that included a description of the proposed evaluation and a budget. The foundation approved our proposal, and our university set up an account and essentially loaned us the money so we could start working on the project immediately. The foundation sent the money to the university a month or two later. On a second proposal, however, the foundation approved the request, the university advanced the money, we started work, but the foundation did not actually pay the grant for nearly a year. Thereafter, the university instructed us not to start work on any project funded by this foundation until money was in hand.

As described in Chapter 8, we established a relationship with a professor of social psychology at a university in the town in which the project would be carried out. He conducted his own empirical research, so he knew the general requirements for data collection and evaluation. He selected some students to be assessors; we trained them, and he supervised them.

Unfortunately, when the project moved to another country, we lost him as a local colleague. We arranged with a faculty member at a university in the next country to take on this role. We kept the same assessors, because it was cheaper to fly them to the new country to conduct assessments than to train a new set of local assessors. But shortly after we started work there, our faculty colleague resigned her role with us, and we were unable to replace her.

Not having a local colleague who could supervise our assessors contributed to another problem. Our grant paid for our expenses, but local assessors and others who worked on the evaluation were paid directly by the foundation. While we met with our assessors on each trip to the country and had a good deal of confidence in them, no one was on site who could determine how much they actually worked. They submitted a bill to the foundation. At one point, the foundation's representative thought they were charging many more hours than it should have taken them to conduct the assessments. We tried to explain that assessing caregivers and children in an agency can require many hours more than the assessments actually take, because the children are not available every hour and the caregiver may not be working that day. However, their fees did seem somewhat excessive, but there was no way for us to know what actually happened or needed to happen. This created considerable friction with the foundation representative.

A Fluid Project

We conducted other work for a new local foundation. While they supported a variety of domestic services for children, we think we were the first foreign institution that they funded. They were still developing their own agenda, and we had to roll with the evolution of their priorities.

In the early stages of exploring what they wanted us to do, we gave a two-day workshop on our previous work improving the development of vulnerable children and how to care for children with disabilities. As a result, they wanted us to provide an extensive training to their professionals on child development, issues of adoption and foster care, and how parents can handle the problems that adopted or foster children often present. They did not want to improve the services their agencies were providing.

We wrote a proposal to do that. After reviewing our document, they cut the project period in half and the budget similarly. Not only did they want us to do everything we proposed in half the time, but in an about face they suddenly wanted us also to help them improve their agencies' services, again with no increase in the budget. After we had developed the materials for the training, they suggested other topics to be included a few weeks before the training was to occur. OK, we can do that. We asked if they wanted an evaluation of the children's developmental improvements as a result of changing the children's services? It would help persuade policy-makers and other funders to do this more broadly in their city and country. They said no.

However, in the middle of the project, they declared they DID want an evaluation of the changes in children's development as a function of the improvements they had made in the agencies. Moreover, it must not cost too much, it must be done quickly, and it must be scientifically respectable. Oh, boy! The timing and amount of money restricted the type of evaluation that could be conducted. The changes had already been started in the agencies, limiting a pre-intervention assessment. Assessing children's cognitive and behavioral development is expensive, and it will take perhaps 2 years to get enough children who have experienced the new services to have enough statistical power to detect effects. And there would be no comparison group.

Fortunately, we had done something like this before in another country using physical growth as an outcome measure and the institution's own data that included before and after the intervention comparisons. They agreed to that. Further, they implemented the changes in services and collected the data; we only advised and consulted on database management and statistical analyses, minimizing a need for a full ethics review.

The point is that some funders can and do make requests for modifications on short notice and even changes in major components of the project at any time. In a very real sense, you are working FOR the funder – they are your partner, and they have the controlling role. You need to do the best you can to work with them to meet their desires. Of course, there can be limits to this process to which both parties may need to accommodate.

10 Implementation

General Lessons Learned

Implementing a project, especially one involving a major research or service intervention in an existing institution with numerous and frequent assessments, can be challenging in any context, but some unusual issues can arise when it is conducted in another country. Such a project may require people to change the way they have executed their jobs for years. Some may not agree with the changes being mandated, some may have training that is inconsistent with the new approach, and major personnel may change mid-project.

The first step in implementing a project in another country is to get to know your international colleagues professionally and personally. This is a common theme in this book because building positive relationships is critical. Many aspects of your colleagues' lives will influence how the project is conducted. They may have training and background experiences that are much different than yours that may lead to disagreements on how to proceed, which measurements to use, and which activities are appropriate. Their religious beliefs may restrict certain procedures, and their family culture and history may hold different values. Engage in social events with your colleagues, not just professional meetings, to get to know them better and to build personal relationships.

Establish a leadership team with your international colleagues. Identify the major players – the people who are necessary to create and implement the project – and meet with this group frequently to design the project. There may be additional people to supplement the leadership team who are needed to implement the project's procedures. This may include someone from your home group who lives on site for a while to help get the project going, or more likely it will be local people who work with the organization in which the project will operate.

Have a professional translator at all your major meetings of the leadership team. Not only is this necessary to understand everything you will need to know, but it can avoid misunderstandings. If possible, it helps to have a translator who is familiar with the subject matter and terminology of your project, for example, research methods, measurements, technical terms, and statistics in the case of a research project. This might be someone from a local university, for example.

Plan to become as familiar as possible with the way things are done and why they are done in that manner before you propose or collaboratively create a project. Some current procedures may seem quite strange, counterproductive, and not consistent with the evidence you know, but usually there are good reasons for them. New procedures may have to accommodate some of these traditional policies, practices, and concerns.

Take steps to engage the major administrators and then the staff of the institutions you are working with and persuade them of the value of trying the new procedures. This may take quite a bit of time and numerous meetings, but it is necessary. Hear their concerns, and take them seriously. People do not perform well if they feel procedures are foisted upon them. Instead, they need to "buy into" the project by participating in its creation and implementation and become genuinely enthused about it. At the same time, there may be one or more who do not agree with it and who may even try to undermine the project. Disagreement and undercutting the project are difficult to determine long distance, especially if your colleagues are not totally frank about these issues during your face-to-face visits. If partners do not feel equal, such meetings will be less forthright, and conditions may worsen in your absence.

To prevent or minimize these concerns, implement some sort of monitoring system and periodic observations to gauge the extent procedures are being implemented appropriately. Expect uneven progress and occasional setbacks. This is where having your own colleague on site can help get the project off on the right foot and alert you to problems along the way that your international colleagues may not think are important or do not want to divulge. Your international colleagues may need an implementation team, which should meet frequently and communicate to you progress and problems of implementation. We also learned to be prepared for major personnel changes during the course of the project. These usually cannot be anticipated or prevented, but don't be surprised if they occur. When they do, stay calm and carry on as best you can.

Recognize that there will be mistakes, data that do not make sense, incompatible computer programs, disagreements over which measures to use, and political and economic events in the country or specific location that can affect how the project operates. For example, there can be international political and economic events between your country and the project's country that can adversely affect the project's budget and produce resentment among team members that stays below the surface but influences decisions and contributes to a lack of enthusiasm to continue to collaborate.

These circumstances can happen in any project, but they are more difficult to handle when you are an ocean away and unfamiliar factors are involved.

Illustrations

Getting Started

There are several steps that are necessary to minimize difficult negotiations and misunderstandings and to maximize authentic application of policies and activities that are mutually agreed upon during the planning stage.

Developing a local leadership team. Changes to an existing organization that are created externally rarely are as effective as those designed with internal partners. Designing a project collaboratively builds ownership of program details, motivates local people to implement project procedures, and instills pride and a willingness to go the extra mile to make the project work. That is why it is critical to develop a local leadership team whose members are the primary "instigators" of change and who will implement or oversee the implementation of the project. This team should be made up of key stakeholders who are respected by those expected to implement the changes in the future.

The project leadership team for one project emerged as a result of our visits and meetings with professionals interested in improving the caregiving environment in one or more agencies. Ultimately it consisted of three international professionals from different backgrounds but who shared a common desire to provide more modern care for young children. Of course, we were included in this team, but we were clearly outsiders at the beginning of the project. That meant we were treated as guests, and disagreements were offered gingerly if at all.

This project leadership team was supplemented in two ways with additional personnel. First, several local professionals assisted the main international members of the leadership team in implementing the project. This larger team was supportive of the changes the intervention proposed and helpful in mentoring staff in one agency. On the other hand, a similar supplementary support team was identified in another agency, which did not work out so well. In that case, there was a major leadership turnover, and some members of this local team were not supportive of the proposed changes.

The second way the leadership team was supplemented was the involvement of an American professional who contributed immensely to the success of the project. We were fortunate to have a US colleague who had a doctorate in early care and education and early intervention for children with special needs, who had directed an early intervention program for hundreds of children in the United States, and was an advisor to the government on policies related to the education of children. Further, she was available to live in the project country for nearly two years to train the specialists and the caregivers in two agencies. This was a fortuitous combination of circumstances for us but an essential component of the project.

Our colleague's first task was to take the suggestions of our international colleagues for topics to be included in the training, blend them with elements of existing training programs in the United States, and create a new training program for caregivers in the project. Then, before she started the training, she spent several weeks with the caregivers and joined them in their daily routine. She changed diapers, fed babies, dressed toddlers for outdoor play, cleaned up after feedings, mopped floors, and smoked with the caregivers on their breaks. She also started to learn the local language, and she actually became rather conversant over the time she was there. In short, she became one of the staff members but ultimately their trainer. She formed personal relationships with the caregivers that were instrumental in gaining their cooperation and accepting guidance from her on how to behave with the children. She was nonthreatening as an observer of their work, their styles, and the actual implementation of the training in new caregiving behaviors.

But this approach also had a professional and personal cost. Our leadership colleagues did not perceive her as the credentialed and experienced childcare professional that she was, and we did not involve her in the leadership team, which reinforced that perception of her. Consequently, while she made great friends with some of the caregivers and staff of the agency, she was not invited to participate in professional activities that her role and status merited – and she was offended by that omission. This talented educator and child professional rightfully considered herself part of the leadership team along with the original five project designers, and neither we nor our international colleagues acted accordingly. During her last week on the job, she disgorged her frustration and disappointment in her notes to us: "…when a handful of people you count on abandon you, God sends you 20 strangers who take good care of you…." These "strangers" were the caregivers (and their families) she was sent to train.

What could have/should have been done to prevent such a disappointing experience by a crucial colleague? In hindsight, we believe expectations should have been discussed early and openly with her and the international leadership team. Professionals often treat other professionals with equality and respect regardless of their assignments, especially as members of a team. This is what we and our American colleague assumed would happen. We suspect she felt we should have included her in the leadership team. We never expected to do that, knowing that every additional team member would complicate the relationships within the team. Further, she was usually busy when leadership team meetings were held. But that situation should have been clarified from the beginning.

In addition, it was possible that the "abandonment" felt by our colleague was due to a culture of hierarchical values stimulated by her intentional affiliation with maintenance workers, laundry and kitchen staff, caregivers, and others who conducted daily menial tasks in their jobs. That affiliation with these workers made her appear to our leadership colleagues as one of them, not a professional doing her job in a very creative and effective manner.

The purpose of her deliberate service with all levels of workers was to understand their workloads and limitations and to get to know each person coming into contact with the children in an attempt to make the training more appropriate and meaningful. In this sense her tenure was very successful. The training was effective, and her personal relationships with the staff often had other benefits. For example, her detailed reports gave us insights that contributed to our understanding of what went on within the agency when other team members were not on site and we were thousands of miles away. For instance, she reported that untrained laundry workers were used as caregivers on days the institution was understaffed, because a scheduled caregiver was moonlighting at another job to make ends meet at home, which evidently was quite common. She also reported that when there was an undercurrent of dissatisfaction by the caregivers which could not be brought up openly in a very hierarchical culture, she was able to sit with them after work hours and discuss strategies that allayed their concerns and plan jointly with them a solution to suggest to the administration without pointing fingers at workers who otherwise felt their jobs would have been in jeopardy.

But this very successful approach to training and mentoring had an undesirable personal cost to her that otherwise might have been avoided, at least to some extent, by a discussion with her and the leadership team of roles, strategies, and expectations at the outset of the project.

Background information and issues. We spent quite a bit of time understanding why the agencies functioned the way they did. A major factor was that the agencies are administered by the Ministry of Health, so medical procedures predominated. For example, each child was given a medical diagnosis. These diagnoses do not necessarily match the definitions of US diagnoses. For example, they still used non-specific diagnoses similar to our now-discarded labels of "failure to thrive" and "minimal brain dysfunction." Nearly all children who arrive at the agency are under-developed for a variety of reasons, many of which were unknown, so they were given these non-specific "diagnoses." Also, since the agencies were administered by the Ministry of Health, every child needed a diagnosis of some sort to justify admission.

Traditional training of caregivers focused on health issues and basic care routines. There was no concept or practice of "behavioral pediatrics" or much concern about early education. There was the belief that children with disabilities could not be cured medically, and therefore nothing could be done to improve their development. While many of these procedures and attitudes seemed strange to us at first, they were understandable given their circumstances and not very different from what existed in the United States a few decades earlier.

The educational backgrounds of some of our international collaborators were also somewhat different from ours. Their psychological training was more psychoanalytic, also similar to what existed in the United States some decades ago. Our orientation was more behavioral. This disparity produced several disagreements over the course of the project. For example, when a child was frequently aggressive to other children, we wanted the caregivers to stop the violence, pay attention to the victim, and implement a time-out procedure for the perpetrator, the procedures for which we wrote out in detail. On the other hand, our colleagues wanted to talk to the aggressor to find out why he behaved in this manner. Our compromise was to stop the aggression, attend to the victim, and remove the aggressor from the situation; then talk to him all you want. It never really happened this way, however.

This academic difference also influenced the selection of caregiver assessments. We wanted to have assessments that measured caregiver job satisfaction, stress and depression, and conservative versus liberal attitudes toward children. Our colleagues agreed, but they also wanted questionnaires about the caregivers' relationships with their own mothers. They thought this would explain the caregivers' behavior with the children. We were more interested in more practical issues, such as the effect of the intervention on the caregivers' attitudes, behaviors, and job satisfaction, and we were concerned that the caregivers would only tolerate a few assessments. Although we all agreed not to add the extra assessment, our colleagues administered it anyway and did not tell us for some months.

We also made some assumptions about our colleagues' experience analyzing data. At the time, Great Britain and the United States were leaders in the development of statistics, and researchers in the United States had vastly more experience conducting empirical studies and analyzing data than scholars in most of the rest of the world. So, we assumed that we and a hired staff statistician would receive the data, put it into a statistical database, and analyze the data from the project. Our international colleagues objected, but only after some time had passed. They did not simply want to send the data to the United States; they wanted a copy of the database that we produced or to create their own database.

This was a gross professional and diplomatic error on our part, and we readily agreed to give them a copy of the database when they asked for it. Over the years, however, it was clear that they and their statistical colleague knew some statistical techniques but had very little experience analyzing large databases, checking on assumptions, and answering substantive research questions. So, we held data analysis tutorials one day of every visit for several years. In the end, many, but not all, of the data analyses were conducted in the United States.

The intervention stressed having the existing special staff members for massage, music, language, and other skills work with children in the group setting rather than pull them out to do it in isolation. This change of practice was intended to permit caregivers to see and follow through on the lessons and exercises. Nevertheless, years into the project, one of our leadership team proudly showed us a videotape of his work with agency children with Down syndrome in which he pulled them out of their rooms and worked one-on-one with them in private. What we all did in the project did not translate into his style of therapeutic care.

Personal relationships. While all this information about our international colleagues and the agencies was extraordinarily valuable, ultimately personal relationships were needed to glue the team together. We attended concerts and other cultural events with our international colleagues, and we held parties in the evening at the agency facilities. A candle was placed in the middle of the floor in a darkened room; music was played; and we danced, talked, laughed, and drank a bit together. Some of the professional staff even tried to teach Bob how to dance – what we do for our project! We tried to learn about our colleagues' personal lives, their families, and their society, but we were steadfast in avoiding anything political. We invited our international colleagues over for dinner, and we went to their apartments for meals. We learned that they viewed this as more casual than we did, occasionally canceling at the last minute. It was all part of getting to know our colleagues and building relationships of trust and friendship with them – at least we thought we did.

Other unexpected behaviors. There were behaviors and events for which we were unprepared. We heard some extreme beliefs or statements from some foreign citizens about the United States. For example, "You Americans adopt our children to sell them for body parts to use in transplant operations." Also, local citizens at that time generally were not sympathetic with adoption. A woman was quoted in the local English-language newspaper as saying, "If we could not have children, I would rather have my husband's brother come over and inseminate me than to rear someone else's child." This was only one woman, and the prevalence of her attitude was unknown, but adoption was not common then. Another apparently cultural difference occurred when we gave caregivers a multiple choice "test" before and after their training to measure what they had learned. We discovered there needed to be a proctor in the room to supervise the testing; otherwise, the caregivers readily and openly helped each other answer the questions.

Also, if some dialogue will be held in the local language, we recommend that projects employ an independent translator, even if one or more of your international colleagues speak English. Two of our colleagues spoke English, but the other did not. At the beginning, we had an independent translator at every meeting who was a psychology graduate student in addition to a professional translator. She would translate for the colleague who did not speak English and for us when our colleagues would hold a discussion in their language among themselves. After a while, we did not have the translator, and our English-speaking colleagues would translate. Occasionally, our colleagues would get into a discussion among themselves in their language that might last several minutes, and then they would summarize it for us in a few sentences. This often did not seem sufficient for us. Chris understood another language that was close to our international colleagues' language. However, she never admitted to understanding some parts of their conversations. One day, however, our international colleagues got into one of their vigorous discussions among themselves in their language over a contentious issue that we all had tried to solve, and their private discourse grew quite animated. We listened to it. Suddenly, Chris burst out, "I did NOT say that." Our international colleagues stopped immediately, looked at us with some amazement, and Chris repeated what she had actually said. But she had blown her cover.

During the course of the project, two directors of agencies in the project were replaced. This was a potential threat to the design of the project. The agencies were not randomly assigned to the intervention and two comparison conditions, but each of the three original directors wanted to be in the research group that they were ultimately assigned. One director died unexpectedly. Her agency was the no-intervention condition, and her replacement accepted that. The departure of the other director was more problematic. She and her main assistant got into a serious disagreement with each other, including accusations of financial mismanagement. It got quite ugly, we were told, and ultimately the Director resigned or was fired, we are not sure which, but the Assistant remained. A new Director had a very different style, which took some adjustment on our part. Fortunately, the Director of the main intervention agency remained steadfast in her position, and without her the project would not have been as successful.

A difficult situation arose several years later when we conducted a follow-up study of children who had been in our three intervention agencies and who were subsequently placed into domestic families. The agency personnel knew these families and how to get in touch with them, but it was difficult for them to do so. One major problem was the "secrecy of adoption" policy, in which adoption was not disclosed to the child or others. We wanted to send an assessor to the adoptive homes, but this was viewed as a possible threat to the secrecy of adoption. We tried to assure parents that the home visitors would not be affiliated with the agency but with the university. Nothing would be mentioned about adoption; we only were interested in the child's development. But even that purpose was potentially threatening. Families might not want to risk the agency (administered by the government) assessing how good a job they were doing as parents.

The agency social workers who were given the task of contacting these parents often knew these families and made judgments about who not to call. If they knew the parent might object or the child was having a problem, they avoided contacting that parent. Some social workers were more willing than others to try to contact parents. This was not the best sampling procedure, but it may have been the best we could get. Sometimes you must settle for the best obtainable, if not the best, information you can get.

Supervision. A major component of the intervention in the agency was the training of the caregivers. We created a written curriculum based on existing curricula and best practices, but this had to be taught to our colleagues and agency specialists, and then they trained the caregivers. In addition, research indicated that training alone rarely produces the changes in caregiver behavior we desired unless it is accompanied by supervision. But supervision, as we knew it, was nearly unknown in the project's country.

For example, when we held our first meeting with agency professionals and staff to explain supervision, one asked us why supervision of caregivers was needed. Isn't training them enough, they asked? Or did we actually mean "control" of the caregivers' behaviors? Caregivers viewed supervision as "spying on them," and they worried that their mistakes would be reported and used to penalize them or even to fire them!

We had to describe, explain, and review over and over our concept of "reflective supervision." This approach conceives of the supervisor and caregiver as equal members of a team, each helping the other to improve their job performance. This concept was inconceivable in this culture, because it put the supervisor on par with the caregiver being supervised. In their limited experience, supervisors corrected staff's mistakes and told them what to do. But we encouraged all discussions between supervisor and caregiver to involve the encouragement of positive ideas and behaviors, and these conversations were to be respectful and emphasize guidance, not punishment or negative reporting. They would begin with the identification of positive behaviors, praise, and encouragement, and then perhaps a discussion of something that could be improved. It became obvious that agency professionals and university faculty needed to be taught how to observe caregivers and how to emphasize the appropriate positive behaviors to be encouraged and limit criticism.

Yet, if supervision and coaching were not recognized methods at the start of this work, systematic monitoring was even more troublesome. We needed to impart the importance of these strategies and how they could be implemented in a supportive rather than punitive manner between caregiver and monitor/supervisor. This needed a delicate balance, so Chris had to play the role of the mentor in demonstration sessions at each of our visits. Critical to the success of these sessions was a system of notetaking and post-session discussions between the observer and the observed. Chris and the lead supervisor would discuss what behaviors were positive and beneficial to the children and one or two behaviors that could be improved at each observation with the observed caregiver. Then, those issues were reviewed at the next visit, which was likely to be at least three months later. Staff appreciated that the discussions were remembered, that they were listened to, and especially that each report began with what they did well.

As described earlier, we were fortunate to have a US professional colleague who was very experienced in training, supervising, and mentoring and who lived on site. She provided examples of monitoring and supervision for the professional staff who would become supervisors as well as the caregivers. Because she had worked with the caregivers, she was nonthreatening as an observer of their work, their styles, and the actual implementation of the training in new caregiving behaviors. This natural give and take between our colleague and the staff made the "reflective supervision" style impactful and comfortable. It also helped the agency professionals and faculty observe supervision in action!! We believe having her there and developing these relationships were essential to the program's success.

Another strategy to encourage and monitor implementation was that our leadership colleagues held weekly meetings with their local team, many of whom were supervisors of caregivers, at which they discussed issues the caregivers were having implementing the project. Many of these meetings focused on a particular child who was having a problem or who was being a problem to other children or to the caregivers, but other meetings concerned caregiver–child interactions that were at the heart of the intervention. Our leadership colleagues wrote summaries of these meetings and emailed them to us to keep us informed of the problems and successes of project implementation.

Training sessions and challenges working with children. The training part of the intervention consisted of first training the agency professionals who, along with our US colleague, would then train the caregivers. The professionals could be trained largely with the written curriculum that our US colleague had produced. The professionals then provided feedback on the training materials and behaviors to be encouraged so the training would be more culturally sensitive and appropriate.

Training the caregivers had to proceed with other techniques. The written curriculum is necessary, but research shows that reading texts does not get caregivers to change their behaviors. There is no substitute for hands-on demonstrations for people who must do hands-on behaviors, especially with children. For example, the standard practice in the agency for feeding solids to infants was for the caregiver to hold the infant on her lap with the infant facing away from her. Then, food was "shoveled" into the baby's mouth with a big spoon, literally one spoonful every five seconds. Each spoonful was followed by two swipes back into the child's mouth of food that was dribbling down the child's lips and face. There was no talking, no eye contact, no interaction except putting food in the infant's mouth. It was very "efficient," especially if you had 10–12 infants to feed in an hour or so. But if you were not used to seeing this practice, it was nearly inhumane, if not "abusive." It certainly did not promote engaged, sensitive, responsive caregiver–child interaction.

To bring these points home to the caregivers in a training session, the trainer "fed" one of our leadership colleagues in this manner. He leaned back in a chair and had a large bib around his neck and over his chest, and the trainer fed him big spoons of food very rapidly, just as caregivers do with the infants. The demonstration was both hilarious and disgusting. Our colleague's face and bib were smeared with food, and he was grateful for the end of the demonstration. He gasped and then comically asked, "Can I have some more food please?" When asked how it felt, he admitted he was almost choking with the amount of food in his mouth at one time, and that he nearly threw up at one point. Well, this impressed the caregivers in a way no text narrative could have accomplished. Suddenly, they agreed they could not feed the infants this way any longer. Mealtime, even for infants, would be different from now on.

Good training for caregivers also requires that the trainers – including Chris and Bob – mentor the caregivers by joining them in interacting with the children. This meant getting down on the floor with caregivers and children – children with and without disabilities, children with or without dirty diapers, children drooling or with runny noses, whatever. We did this at every visit to demonstrate what was considered best early intervention practices and to support the caregivers. We were also very aware that the children were within earshot of all comments made between mentor (including us) and the caregivers. So, in addition to focusing on the skill being demonstrated, how to talk about a child in front of the child had to be carefully demonstrated and managed.

After returning to the United States frequently with colds, sniffles, rashes, and other maladies, we quickly realized we needed to have more than the typical set of immunizations, including hepatitis A and B. We also moved to demonstrating a variety of child development techniques on dolls and other adults, instead of the children, similar to the feeding demonstration described above. These often turned out to be loud and raucous affairs that were fun and promoted social bonding. We tried dressing another adult who imitated some of the children with disabilities, that is, with certain spastic or muscular contortions and the inability to move their arms or legs. We also had caregivers in the training sessions stand in the corner for a long period of time or sit in awkward positions in a chair to help them understand what it is like for a child with disabilities to be ignored or not positioned properly so they were comfortable and part of the group interaction. Caregivers reported that these demonstrations were fun and "hit home." The next time the caregiver or therapist was with a child and started to reflexively repeat old habits, they suddenly remembered to change their behaviors. This did not supplant getting down on the floor with a child at the time the caregiver had a question, but it did wonders to illustrate new behaviors and make the training both instructive and entertaining.

Preparing for other changes. In addition to training caregivers, our intended intervention was to make the agency structurally more family-like. This meant creating smaller groups of children with their own playroom and dining area. But the agency was built for larger groups of children, perhaps 10–12, so rooms would need to be divided and some substantially reconfigured. To plan for such renovations, we and our international colleagues needed to understand the building and its current structure.

This began with a discussion of the information we needed to gather and the types of documents that would be helpful. Then, we all took a walking tour of the facility that included following building blueprints, a new activity for some of us. However, this was very important to designing new rooms and to limiting the size of groups. We needed to know which walls could or could not be moved, where the plumbing was located because that could not be changed, which walls were soundproof for sleeping, and even on what side the sun came up so we did not put the sleeping children on that side! Of course, ultimately architects and builders need to ensure the accuracy, feasibility, cost, and legality of our speculations. We cannot emphasize enough how gathering this kind of detailed information required constant and high-quality translators and flexibility and accommodation between what we wanted and what was possible.

Commitment and desire for change. International projects must include commitment and desire for change by many local and external stakeholders, including policymakers, practitioners, educators, and others who are necessary to implement the intervention. It takes sensitivity, timing, and a partnership style for such commitment to develop. In some cases, the desire for change is initiated by the international partners and the external collaborators are brought in to assist. In other cases, the US partners are the initiators of the project, but the project will not succeed without the commitment of the local partners. In either case, commitment and desire for change must exist on both sides and ideally by all who are responsible for implementation.

Building commitment in international colleagues can take many forms. The first people who need to be committed are the main administrators of the organization in which the intervention will be implemented. Fortunately, the Director of the main agency in our project was part of the leadership team and was enthusiastic about all the changes from the beginning. She endorsed the changes the leadership team decided upon and made it clear to staff that these changes were going to be made. If there was resistance, she would insist that "we give this a try for a while." She was also instrumental in getting the directors of two other agencies to participate.

The leadership team, and especially the Director of the agency, held several meetings with professional and caregiving staff once they decided the nature of the changes they wanted to put into place. These meetings were conducted with a blend of "here is what we had in mind and why we should do this" on the one hand with "how do you feel about these changes and what would be difficult for you to implement them." Ideally, the leadership team might present the need for changes and what they hope the changes will accomplish, and then ask staff how they think they could accomplish those outcomes. The task is to help the staff's vision converge on the leadership team's vision, which is presumably based on some research evidence. Alternatively, the leadership team can present the elements of changes they have in mind and ask for staff opinions and concerns. We did the latter.

Staff had concerns about how things would work. Caregivers who had never worked with children with disabilities wondered how they would deal with such children in their group. They were assured that they would be provided training and support. Another caregiver asked a more conceptual question: "If we are changing our behaviors so that the children develop a relationship with us, won't they have a more difficult time when they leave the agency, many of whom go to families and need to make new attachments, and some go to another service where there will be no relationships?" Bob said he thought that they would be better off in both cases, but that he would consult with several of the world's authorities on relationships and attachment and report the answer at the next meeting. He did email this question to three world specialists in attachment. They all said direct research evidence was lacking, but existing evidence and theory suggest that the children would be better than not having such experiences, which is what they have now. At the next staff meeting, Bob reported their opinion and concluded, "It is probably better for them to have 'loved' and even then lost than to never have 'loved' at all."

Sometimes promoting commitment can be more subtle. For example, agency staff are very accustomed to visitors coming in and out of their rooms. Some just observe while others can be more intrusive. In our case, it was important that they perceived us as being committed to this work for the long haul. So, although it was time consuming and expensive, our return to the project site every three to four months over the years was one such demonstration of our commitment to the project on which we were all working.

This persistent presence was not lost on the agency staff. As one professional reported hearing in the staff lunchroom: "Did you see those two Americans again?" a relatively new caregiver remarked.

"Yeah, Bob and Chris have been here many times and even know some of our names and some of the children's names, too," answered a long-time laundry worker. The professional pointed out to the group of staff members that if the Americans were that committed to helping the children who lived there, just think how committed the workers should be!!! News of this conversation spread to many of the other workers, and reports indicated that it improved the attitudes of a few staff who were initially less enthused about the project.

A Different Project

Another project was instigated, funded, and directed by the Head of a new foundation that was just starting to improve caregiving in low-resource countries. A major issue for us was knowing precisely what our role was, if any, beyond being the evaluator of whatever project eventually was implemented. We were asked occasionally about what should be done –and occasionally we volunteered our thoughts – but our role beyond evaluation was ambiguous and sometimes delicate and should have been more clearly established from the beginning.

Similar issues of role and responsibility arose in other domains. The Foundation Director had been working in one country for some time before we arrived. She took a rather "top-down" orientation, preferring to start with getting the blessing and endorsement of the President of the country and other major ministers and then, with their backing, getting institution directors to agree to implement changes. There was a good rationale for this approach, we eventually learned. In some countries, the President and a few families literally "ran the country" and not much happened unless they were behind it. She was very good at getting meetings with the President and other relevant officials and soliciting their approval, despite the delicate fact that if the country needed to improve its services for children then the current services were not of very good quality, which could be embarrassing to the President. Our experience and orientation were a bit more bottom-up, trying to convince the agency's Director that things could be better and having her appeal up the hierarchy for change. Ultimately, both approaches are needed, but working them out and balancing them could be difficult.

At the onset, we attempted to get to know our on-site local colleagues who worked in the agencies and how things were organized. This was a bit challenging, because we seldom had professional translators available to us. Once again, we cannot emphasize enough the need for excellent professional translators when working in a foreign country.

Also, as the evaluators, we came into the project later than the trainers and interventionists. Therefore, many assumptions were made regarding what we should have known about the cultures of these countries. And our responsibilities as evaluators were, at times, expanded to include training and mentoring. All of these circumstances were understandable but made for less than smooth sailing in our collaborations.

Developing an international team. At first, the culture seemed very deferential to professionals from other countries. Recognizing that, we needed several partners from the country in which we were to work. But we felt we were climbing uphill to gain enough trust from these partners so that they were willing to tell us what was needed and why. However, soon we learned that what we thought of as simple cultural differences were complicated with differences in religious tradition, a sense of nationalism, a history of American political intrusion, and gender inequality. We accepted that we were not there to change these differences but rather to work within them.

The first step in working within these differences was to establish with our international partners that the project was a collaboration partly facilitated by us but driven by them. While the Foundation Director and we had definite ideas about what might be done to improve the services, we had to convince the international partners that we were not there to drop a prefabricated intervention into their agencies. They had to help create it. We had to emphasize the importance of open and frequent communications, that we knew they understood the needs of the agencies better than we did, and that there was no right or wrong during this planning process.

Our progress in this endeavor was greatly enhanced when we were fortunate to gain the employ of an English-speaking local Jesuit priest who was also a university faculty member in the social psychology of children and youth and was trained in the United States. The nuns who directed the agencies were familiar with him and approved the project whole heartedly once he was named. It was not clear to us if this total acceptance was based on this particular priest's personality, talents, training, and skills; his position in the Catholic Church; the fact that he was a man; or all of these. It did not matter to us; we were grateful for his involvement. Once again, having professionals from within the country is essential.

Although we did not have a US colleague living on site, our priest/ university liaison played that role. He was the best combination of someone living in the culture and yet understanding ours. He was accepted by the internal and external key stakeholders as a qualified and appropriate representative of their political, academic, and religious communities. He hired data collectors and trained them and became their confidant so that internal issues were communicated to him early in their development and potential problems were averted.

And he solved problems. For example, a data collector, Joanna, voiced a problem. "Father, I cannot use the project's laptop to collect data at each of the two agencies, as you requested."

"But, Joanna, we bought your laptop for that purpose!" exclaimed Father G.

"I know, but I am afraid to carry it with me to the bus stop because I may be mugged, and it will be stolen by gang members!!" Father G. realized she was right. Together, they came up with a plan to request funds for an additional computer and storage cabinets so that they could store and lock a PC in each institution and enter and transfer data only while inside the building. Crisis averted.

Similar issues concerned not paying caregivers on Fridays and not having them travel to work in the dark. These were issues that would not have occurred to us, but our local professional vouched for their veracity and relevance and helped the agency formulate new policies to avoid these risks.

Another International Colleague

In another project, another country, and on another continent, the value of having a local colleague to help us understand cultural issues and to implement certain procedures was again demonstrated. In this case, we had several projects to implement and different project roles, and we were fortunate to have both an American colleague born in the country of the project lead our delegation who engaged the cooperation of a well-known, exceptionally well-liked, and very well-connected international faculty member. Through his work we had "ready-made" local teams of willing non-profit (NGO) directors, academics, and students to be mentored in interventions, data collection, and analyses. This arrangement was unusual and extraordinarily beneficial for us, because many of the typical roads to a successful international collaboration were paved for us by these two colleagues with minimal to no effort on our part.

The initial steps were to get to know our other local international colleagues through a series of meetings, professional and social, and to do so over time. The agenda for these meetings included reviewing what we had done in other countries and to discuss how those efforts could be adapted to accomplish goals of the cities in which our local colleague and his partners were interested. He had his own local team, and the members already had a commitment and desire for the changes we proposed. We found many areas of knowledge and practice that overlapped with ours and critical values that we agreed upon. He then took the lead at all meetings we had with institutional directors and government officials who were critical to the success of our partnership. These were viewed immediately as locally created and locally led projects supported by American technical skills and experience.

Our international professional colleague also knew other academics and social welfare professionals as well as the relevant government ministers. This led to a collaborative publication in an international journal as well as meetings with top government officials who saw the potential for broader application of our work across the country. Although these expanded ideas dissipated when the government officials left office, they illustrated the range of potential activities when prominent local professionals are involved from the beginning in collaborative projects.

Yet Another Project

In this case, the project was not identified at the start of our work, which began with a series of visits on our part to hospitals and NGOs as well as meetings with government officials and other powerful decision-makers. These were arranged by our international funders, a newly created foundation managed by prominent citizens. We were invited to give lectures and media interviews before the local stakeholders asked us to help design and implement a project of training and changes for some service agencies. The purpose of these improvements was to help children of all ages to be more "adoptable" and to better prepare families to adopt. In addition, they recognized a need to develop an in-country team of professionals who could strengthen institutions in transition and support families with adopted or fostered children.

Developing a local team. Our local sponsors asked us to put together a curriculum and a team of international trainers while they selected local in-country members to become a team of train-the-trainers to be sent throughout their country. This group helped us to understand their priorities and shape the training materials to address those needs.

A residential colleague. We had two professional colleagues to help us. One was a colleague who worked with us on previous projects who actually was born and raised in the new country. He understood the culture and had years of experience with the goals of our training and our partnership style. He and his very experienced colleague were the primary contacts for monitoring this work on-site. They were very sensitive to the implementation issues within the agencies, participated in the training we jointly provided, and moved at a pace comfortable for the local team members.

We also engaged a professional member of the sponsoring foundation who became our principle local contact for conducting a partial evaluation of the train-the-trainer intervention, which was then implemented by her and her colleagues in the service agencies. She had relevant professional training and was well-prepared to collect data on children, keep records of the training of local professionals and caregivers as well as the interventions that were implemented, and bring the report to the foundation and policymakers. There is no substitute for local colleagues who will actually implement the collaborative project.

Part III

Issues, Crises, Threats

Collaborations, almost by definition, are potential breeding grounds for conflicts. A good collaboration includes partners who possess complementary skills and responsibilities each of which is necessary to the success of the project. Therefore, the partners are likely to have different attitudes, perceptions, opinions, and values that may evolve or erupt into conflict at times. Further, because each partner is necessary for the project to succeed, each of these differences must be dealt with and resolved.

This is why we have suggested that you get to know as much as possible about the backgrounds of the people with whom you will work closely and the organizations that will be involved in the project (see Chapters 4 and 10). Even then, conflicts can arise at any time, and most cannot be anticipated.

Consult with all the people who are relevant to the project, and the bigger the project, the more people who need to cooperate with it. Further, the more administrative structures that govern its implementation and existence, the greater the potential for disagreements and problems involving project participants and stakeholders outside the project. Getting as many people in authority who have some potential interest in the project to approve it at the outset is advisable, but it does not assure smooth sailing over the entire voyage. People change their minds, circumstances change, and people leave their positions and are replaced by individuals who personally did not approve the project. There may even be accusations, true or fabricated, designed to impugn the integrity of the collaborators and shut the project down. The motives may be legitimate professional and administrative disagreements or personal animosity, jealousy, and self-interest. For the most part, these cannot be anticipated, but they must be dealt with by you and your colleagues, which can be difficult when they are occurring in a different country.

Large projects involving many diverse stakeholders, especially those projects that will operate in the community and involve community agencies, have greater potential to create problems. Bob, a veteran of traditional domestic lab-based research, observed, "I've never been in projects in which there are more people who could shut the project down in an instant for legitimate or illegitimate reasons." Chris, who had managed numerous domestic community-based collaborations, was accustomed to this situation. She knew the importance of getting all the stakeholders on board and cultivating and maintaining good personal and professional relationships with everyone. This helps, but it is not necessarily sufficient.

In Part III, we describe several major issues that have the potential to produce issues, problems, and conflicts.

11 Unexpected Intervention Collaterals

General Lessons Learned

Expect extraneous factors to influence your project. Interventions conducted in real community settings typically consist of a collection of changes implemented more-or-less simultaneously as a package. Given that the full intervention consists of such a collection of changes, it is usually impossible to determine the contribution of each specific change to the total result. This leaves deciding what parts of the intervention contributed most to the outcome to inference and speculation. But those elements that made the intervention work or not work are likely crucial to your interpretation of the results.

Do your best to identify "collateral" influences. You are also likely to suspect that certain components and characteristics – we call them "intervention collaterals" – that were indirectly produced by the intervention were important contributors to the outcome. These are characteristics that were not deliberately or directly changed or measured that the intervention produced that likely contributed to some of the effects that you attribute to the intervention. For example, perhaps the Director of the agency in which the intervention was implemented was thoroughly committed to the project's changes and insisted that his staff implement them. Without his on-the-ground leadership, you doubt the intervention would have had the substantial effect that was demonstrated. His commitment and enthusiasm were not something that you manipulated or deliberately promoted but they were created by the intervention – they were intervention collaterals that helped produce the end result.

Similarly, it is possible that all the changes implemented as part of the intervention produced an attitude among the staff of that agency that major changes were taking place, which likely increased their motivation to implement the changes and created an expectation that the children would improve. That aura of expectation for success was not created in a no-treatment control agency. The enthusiasm in the full intervention versus the relative lack of enthusiasm in the no-treatment control agency was another potential intervention collateral.

It is very important to be mindful of intervention collaterals that both support or impede the effectiveness of an intervention. Because collaterals are produced by the intervention, identifying them helps to explain why the intervention worked or did not work. They are not "confounds," which are correlates of an intervention that the intervention did NOT produce yet contributed to the result. It's important to keep this distinction in mind, even though the line between the two can be blurred.

Illustration

Suppose a large, complex intervention consisting of many changes in the services provided to children was implemented successfully in one agency whereas only training of staff but no other changes were made in another agency. The results showed that the intervention produced improved caregiver services to children and the children displayed improvements in their physical and social development whereas those improvements did not occur in the comparison agency. After the project was finished, consider the following conversation among the project leaders.

"There just was not the enthusiasm and commitment to change in the comparison agency as there was in the full intervention agency," reported Samuel, who spent time each week encouraging the caregivers and mentoring the professionals to supervise the caregivers. "In the intervention agency, everything was changed, and the changes were visible every minute of every day. All the caregivers knew they were a part of something big. But in the comparison agency, there was no physical sign of changes, so they didn't change their behavior."

"They did not have the commitment of the leadership either," observed Chris. "The Director in the comparison agency did not encourage his staff to implement the training they received. The staff never bought into the training, even though the Director stated at the outset that he wanted to be in the training condition."

"I think much of this stemmed from their reputation," Samuel offered. "When we started, this agency and its staff had the best reputation in the city. They had a new and spacious building, their teachers were regarded as being the best in the city, and their Director was sort of the 'Dean of Directors' in the area. They felt they knew it all and did not need additional training. And in one way, they didn't. If I asked them what they would do in one or another situation, they would give the right answer verbally – but I would not see them actually do it with the children. They would go on doing what they always had done – they were 'the best' – and the Director did not encourage them to perform any differently. Had the Director encouraged them to implement the training, they might have done better."

"The training tried to encourage the caregivers to form relationships with the children, but they ignored this because it conflicted with their usual behavior," speculated Mary. "They don't want to interact with children in a personal and social-emotional way; they don't want to get close to children because they soon leave the agency. So, they continued with their impersonal behavior toward children."

"In addition, they followed the old, traditional approach that these agencies have followed for decades," continued Mary. "This works against us in several ways. First, the administrators of their district insist on the traditional medical orientation, so the Agency Director follows their lead and doesn't want to resist them. Second, the pediatricians and medical nurses have the power in the agency while the educational/behavioral/ teaching caregivers are treated like second-class professionals. The medical people do not want the educational/behavioral/teaching caregivers supervising the medical nurses, so they did not insist on them implementing the behaviors we taught them. The medical people have no use for social-emotional caregiving – they have no idea that it might improve children's development, including their physical development. When we showed them the graphs of how the intervention children improved in physical growth, they said, 'You made up those data.'"

"In contrast, all of the changes in the intervention convinced those staff members that their behavior mattered; otherwise, we would not go through all those changes," Samuel reasoned.

"Also," added Mary, "we renamed caregivers primary and secondary instead of medical and non-medical teachers in the intervention agency, and we did a good deal of team building. Both of these components helped to dissolve the distinction between medical and educational/behavioral personnel. But the comparison agency still had that old distinction and status difference, and it perpetuated the lack of emphasis on the quality of interaction with children."

"Bob and I would meet with the Director and Pediatrician on each visit," Chris remembered. "And they would tell us that they were implementing everything and that the caregivers and children were improving. But when we visited the wards, not much had changed. We would tell them about a few good caregiver–child interactions we saw to encourage the good behaviors, but we also described to them all the old less favorable behaviors we saw. It did not seem to make much difference – they talked a good race but did not run one."

"It may be hard for them to see what is wrong on their wards," thought Mary. "I remember Bob saying to Chris when he first came here that he didn't see anything that was so bad about how the children were treated. And Chris told him, 'the problem is what you DON'T SEE.' It is difficult to see what is missing. If you are used to seeing it all the time, it looks OK."

"And when we showed the Director and the Pediatrician the graph of data that demonstrated that their children were not improving very much compared to the full intervention children, they were shocked," said Bob.

"They thought they were the best, and now they were not. The torch had passed to another agency! But instead of motivating them to improve and implement the training, they denied the facts and stonewalled the training. Training alone, as has been shown in other contexts, just does not change people's behavior very much, and we have seen several reasons why."

12 Train Wreck Coming

General Lessons Learned

Projects can face a variety of threats to their integrity and even to their existence. This chapter recounts two kinds of threats. One is starting a new project in which a variety of challenges are known and recognized as serious difficulties from the beginning. The other kind includes unanticipated events that present difficulties for the implementation and nature of the project.

Do not start a project if you think the necessary ingredients are not or will not be available or achievable. This seems obvious, but sometimes the situation appears to command that you conduct a project anyway.

Every project has several components necessary for its success. Some of the major ingredients include sufficient funds to support the project, a planned intervention that has a reasonable chance of being successfully implemented, and certain skilled and committed personnel to implement it. Not all of these components may be known at the outset. For example, you have to plan the intervention before you can apply for money, and people may need to be recruited and trained as part of the project. But try to monitor the essentials as planning progresses and have the courage to stop a project when it becomes clear that one or more crucial elements cannot be achieved. Of course, the purpose of most research and service intervention projects, for example, is to determine if the intervention or service will be effective, so the outcome cannot be assured, but you need to have confidence that the intervention or service can be faithfully and competently implemented.

Pursuing a project for ethical reasons or because it seems the responsible thing to do are worthy motives, but they should not lead you to pursue a project when major ingredients are not available. For example, it is ethically responsible to provide comparison groups that do not receive a successful intervention as part of the project to get that intervention after the project is completed. But once the project has demonstrated the intervention works, it is difficult to convince a granting agency to replicate the project unless something new will also be tested.

Alternatively, a previous funder may want you to conduct a project they are willing to fund but at a level that is inadequate to accomplish its goals. It can be difficult to "advise" the funder that the project cannot be accomplished for the amount of money being offered. You may feel it best to do the funder a favor and "give it a try," but it is no favor to waste grant money on a project that you are confident is likely to fail. Sometimes it is necessary to provide the funder with your "professional advice" not to do the project or try to work out a compromise with less ambitious goals that has a better chance of success. The decision can be agonizingly difficult and the outcome not clear until you have actually given it a try.

If your project requires that local staff change the way they perform their jobs, plan on a substantial amount of time to convince them the changes are beneficial and to train them to perform differently. Most staff of existing organizations resist changing the way they do their jobs, so you can expect some push back to changes you want them to implement. It helps if the senior administrators are thoroughly committed to the project, but even then, staff may need persuading. You cannot force the changes on them, you have to convince them the changes will benefit them and others. Sometimes this task can be very challenging.

Even if your project is well-planned, funded, and implemented, "stuff" happens along the way to completion. Many projects operated internationally and collaboratively take advantage of different or unique groups of people, organizations, or circumstances. This may be the main reason you engage in these projects. In essence, then, you are likely not in total control of all the factors that can influence the operation and outcome of the project.

Most projects require the continuing cooperation of numerous people, especially those projects operated within existing organizations. They may be subject to funding, political, economic, personal, and other factors more-or-less beyond your control, including pandemics. Societies, organizations, and individual people do not stand still for the several years your project operates. Things happen during the project that are unanticipated and potentially threaten the operation and outcome of the project. You need to expect these bumps and hurdles in the road and cope with them the best you can.

Illustrations

A Mountain Too High

Consider the following situation. You have completed an intervention project which was successful at implementing changes in an agency, including improving caregiver behavior with the children and improving the development of vulnerable children. Suppose there was a comparison agency that did not get the full intervention but did get some staff training, but their children did not improve developmentally. Your funder now wants you to give the intervention to the comparison agency but is only offering a small amount of money to do the job. Further, there are reasons why this agency might not be willing and able to implement the changes. Do you give it a try despite these obstacles because it is ethically the "right thing to do," or do you advise the funder that in your opinion this new project could not be accomplished successfully? What follows is a description of what the planning and implementation process might look like and the hurdles, roadblocks, and difficulties that could be encountered on the way.

From the beginning, it was recognized that all the ingredients that contribute to the success of the original project were not available or not in place at the start of this extension. Specifically:

- The "grant" was for a very limited amount of money and for only three months; more money after the initial three months was intended but uncertain. This meant that only one or two units of the agency could possibly be modified, not the entire agency. This partial implementation had the potential to produce schisms and rivalries within the agency.
- The administrative staff of the agency were not enthused about implementing the changes that were needed. In addition to the senior staff, regular caregivers were resistant to the proposed changes. This is not uncommon because most staff resist change, but we would have very little time to get them "on board." This was complicated by an extremely serious schism between medical and educational staff who would have to work as a team in the new system.
- Our policy of paying staff extra from the grant for extra work (e.g., working more hours in a week) or for performing different work from their usual duties meant that some staff would be paid more than other staff by the project. This created dissention, because staff perceived the money as "participation payments," which should be equal, rather than "payments for overtime work."

Therefore, we saw this train wreck coming, but we felt a duty – perhaps a personal and ethical obligation – to give it a try. However, our advice to you, the reader, is not to begin a project with this many challenges, limited resources, and inadequate time to resolve them. What follows is meant primarily to illustrate how an intervention project might get planned with community partners and the problems the above limitations pose, and why one should *not* embark on a project under these circumstances.

Time and money. Chris began the discussion among our major international colleagues from the previous successful project. "We have a commitment from one of our previous funders for a specific amount of money to implement the changes we made in the previous agency in the comparison agency over a period of three months. The funder recognizes that it will take longer than that, and she has the intention of contributing more money over a longer period of time to complete the process, but she cannot guarantee that at this time.

"Frankly, I think we need to do the best we can with this amount of time and money and try to plan to conserve funds and to use money from other sources to the extent possible. The funder, who has helped all of us over the years, is recovering from a serious health problem, and her agency is not doing much business while she recuperates. This means the funding organization lacks her leadership at this time and has less income. So, the likelihood of more money is uncertain at best."

"We have told her," continued Chris, "that it took the whole first year to implement the intervention in the original project, not three months. I think she understands that, but she may not have that much patience. She wants to fund changing the entire additional agency, not just two groups, but she cannot contribute that much money right now."

"Which unit will receive the money and administer the grant?" Ellen asked. "The original grant we had was given to your US university, which subcontracted with the agency, and Charles and I administered it. But these groups are not involved in this project."

"Maybe the money should go directly to the new agency that will implement the project?" wondered Sara.

"I don't think so," countered Charles, who was Director of the first agency. "The agency has no experience managing a grant like this, and the Director is new. Also, many unusual personnel issues arise, and you need to make some tough decisions; some caregivers get upset. I had to face all these conflicts for the first time, and it is very difficult for anybody. Besides, the grant would pay the Director to oversee implementation of the project, and the Director of this new agency does not want her city administration to know that because they do not think she should be paid extra."

"I think we should use the administrative arrangements we had for the original grant," suggested Chris.

"But none of those units is involved in this project," objected Ellen. "Your university and the original agency are not involved. But Sara and I are participating and will get paid for training and supervision, so my university should receive and administer the grant. It does not like its faculty to work on grants that are not held by the university, because it wants to show that its faculty are getting grants, and we get professional credit for getting them too."

"I understand these issues, but I think we need to step back a bit," suggested Chris. "The funder will want to give the grant to our university. It is politically better for her money to go to the university. Also, there are many legal protections for a grantor provided by the university in case something goes wrong.

"In addition, our university was willing to write a contract with the original agency that was simple and direct, without all the legal wording that such contracts often contain. I suspect that your local university would want all kinds of provisions in a contract that we do not want to deal with."

"Also," Chris went on, "Charles will take a major role in administering this project and help the Director deal with the issues that come up. I suggest we stay with the group we have worked with on the first project."

"So," summarized Bob, "the funder gives the money to our university, which writes a contract to the original agency similar to the previous one, and that agency writes a contract with the second agency to implement the project. Charles and Ellen have the authority to allocate money to the second agency for specific expenses as before."

"We will also need a work plan," added Chris, "that gives a step-by-step sequence of actions for the new agency to implement. And we need to motivate them, so we should pay them only after each step is completed, a small amount of money for each of the first steps and progressively larger amounts for each subsequent step."

An unenthused administration. "Alice, the Director of the agency that is to make the changes," observed Chris, "is not Charles. She says the right things when we visit, and she certainly gives us a great deal of respect. She always wants to know what 'Chris and Bob think.' But she seems to believe things are going well in her agency when they are not."

"Yes," agreed Bob. "At the start of each visit, she tells us how well things are progressing. Then, we visit the wards and report back to her. Of course, we try to describe some good caregiver behaviors we observed to identify and encourage the behaviors we want. Then, we have always said what needed improvement, which usually was quite a bit. Maybe she only hears the good stuff."

"People say she really doesn't spend much time in the groups of children," reported Charles.

"I just think the administration is not motivated or committed to this project yet," guessed Bob. "For example, they are already saying they can't divide some rooms to house smaller groups of children. It's too much money or they don't have enough space, they argue. I don't think they have tried hard enough. This agency is twice as big as the original agency; they have all kinds of space. I'll bet the five of us could design it.

"Has she visited the original agency to see how it operates?"

"Yes, many times with her in-house doctor," said Charles. "But they miss the point. She focuses on the physical environment, and, of course, her agency is much bigger and more modern, and the doctor notices the new washing machine. Other Directors come in and immediately say, 'Oh, this is very different. The caregivers and kids are having a good time together, there is laughing and noisy play. It's like a big family.' Not them! They don't understand why the children are better."

"Maybe someone needs to point out to them all the specific behaviors that caregivers do that caregivers in their agency do not do," suggested Bob. "When I first came here, I did not see what was missing in these agencies. They need to see the contrast."

"They have seen it, and it's in the training," said Sara, "but they often don't come to the training."

"Alice, the Director, is taking everything slowly and step-by-step" reported Charles. "As you know, there were some administrative problems in that agency before she came, so she is being especially cautious. That is also why she is so careful to obey the requests of the city administrators and to follow the rule book to the letter. But sometimes, as an Agency Director, you have to do what is right for the children, not the administrators."

"I also don't trust the administrators to tell us what is actually going on," said Bob. "They say they meet regularly with their senior staff about the project, but we know from those staff members that they do not. They say certain changes have already been made, but we can see that they have not. They try to put on a good face for us, but it's not true. And they are always in agreement and positive with us, but we know from other staff that some are against the project. So, what do we do with this situation?"

"We cannot challenge them on these issues," advised Chris.

"Seems like we work with the doctor," advised Ellen.

"She is being more responsible recently," observed Sara.

"Yes, but we cannot make it look like we are going around Alice," cautioned Chris.

"Right" said Charles. "Alice delegates responsibility to her, and she must make sure she is kept informed of everything. She also may come around to supporting this project more if she has some responsibility and control over it."

"And she gets paid for it," mused Sara quietly to herself.

Resistant staff. The new project began with training that was intended to motivate and prepare them to make the new changes. Here is what can happen, and it illustrates the complicated task of trying to train and change a resistant staff.

Ellen began a report on the training of the Director, medical doctors, and other senior administrators and specialists – a total of 15 – from the new agency. "Their training lasted 10 days. The Director did not attend much. We dealt with which groups would get the changes, how the space would be divided, and what staff schedules would be like. This did not go smoothly – there was a great deal of resistance and anger at times.

"On Day 1, the Director introduced the project and generally described what would change. Then, I showed them the results of the original project in which the children had better physical growth and behavioral/cognitive scores than their children. They were shocked. A few did not believe the data; some were almost angry. They thought they were the best service agency in the city. How could this be? They came up with several reasons, but none really applied. Two specialists had visited the original agency and liked it, but they found it hard to believe it made that much difference in the children. Most remained in disbelief.

"On Day 2, Sara and I said we were not their 'teachers' but rather wanted to answer their questions, provide information, and get them to think through these changes. Which aspects of the project did they want to discuss? They started with the issue of which of them should do what. The distinction between the medical staff and the teaching staff erupted immediately. There has been a major schism between these groups long before this project. The medical and teaching staffs even sat separately from each other at this training. They quickly exploded, were very emotional, and demonstrated lots of anger. This rivalry took over. They forgot the question they were supposed to discuss; they just argued. This went on for maybe 25 minutes. We just let them – how do you say – 'vent.'"

"Goodness," exclaimed Bob, "this issue must dominate how they work with each other. It was probably wise that you let them go at it for a while without trying to change it."

"Yes, I think so," agreed Ellen. "Eventually, we said, 'Do you know what you just did? Your emotions are so high and aimed at each other. How can you focus on helping the children?' We let them think about that until the next training day. But for us, it showed how important it would be to write new job descriptions for the medical and teaching staffs, just as we had done in the original agency."

"How did Day 3 go after that?" Chris eagerly asked.

"Well," Sara assured us, "it was calmer – it could have been worse, I suppose. We asked them what they could do to work better together. They offered things like one person talks at a time, people should not interrupt each other and wait until a person is finished talking, they should try to control their emotions, and so forth. More or less obvious suggestions but a good start. We then separated them into two groups to think about dividing a room into two smaller groups of children. We hoped they would try to implement their own suggestions about how to cooperate in these discussions, and they did to some extent."

"On Day 4," Ellen continued, "we had them think about the pros and cons of age and disability integration of children within a group. This, of course, would be a pretty radical change for them, so some resistance was to be expected. We also moved to discussing primary caregivers and their schedules, also a major change for caregiving staff but not the administrators in this specialist group. We focused on principles of care, not the specifics.

"On Day 5, the Lead Doctor presented a plan for scheduling caregivers, which was a step toward implementing changes rather than resisting them. But there were plenty of fears, anxiety.

"By Day 6 there was recognition that new job descriptions for both the medical and teaching staff were needed, and they discussed what they wanted a 'primary caregiver' to be like and whether she should be a medical or a teaching person. Then, we moved to dividing the large rooms into two smaller rooms for children. They were uncertain how this could be done, so the group went to the unit to study how it could be divided. Well, there was not enough space for two groups of children, but eventually both medical and teaching professionals were willing to give up some of their own space to benefit the children. PROGRESS! We told them, 'See, you can solve these problems in a cooperative way.'

"On Day 7, some specialists admitted they did not know how to do certain things. The medical people said they did not know how to educate children, and some specialists did not know how to care for children with disabilities. We assured them that there would be training for those skills, and also there would be supervision, something that is unknown in the agencies.

"So, supervision was the big topic on Day 8. They shared their fears about this project and what they and staff would have to do. They could help each other. Then, what characteristics should a primary caregiver have? Each type of staff thought they were best suited to be primary caregivers, but each soon discovered necessary skills they did not possess.

"On Day 9, we tried to deal with two major limitations in the staff that we have observed. One is that the staff seemed to know the right answers but cannot get themselves to do the behavior with the children, and the other is that the staff are not very sensitive to others and do not express emotions with the children. So, we discussed showing emotions, especially smiling, laughing, praise, sharing, empathy. We emphasized attunement – contact, listening, attentiveness, and so forth.

"Day 10 focused on organization, job descriptions, supervision, roles for specialists, etc. The discussion became quite specific about roles, duties, schedules, etc. They were on their way, but still a long way to go. They agreed to meet as a group every two weeks."

"OK," said Chris. "These were the administrators and senior staff and specialists. Next was the training of the caregiving staff. How did that go?"

"About the same," reported Sara. "The caregiver staff training was intense and lasted over a week for three separate groups of caregivers. While there was some overlap in these groups, generally the training for caregivers covered different topics than the training for administrators, focusing more on the quality of interactions with the children. The administrators and senior medical people were invited and encouraged to attend, but they came to the first session or two and that was it.

"After the project was described and the positive results for the children in the original project were presented, we got the same angry resistance we had gotten from the administrators. Despite the data on the benefits for children, they did not understand how these changes would improve the children. They rejected the results, and one said, 'Anyone can draw graphs,' and another shouted, 'The data be damned.'

"Also they immediately raised difficulties in implementing the changes and did not understand why working five days per week was better than their current schedule. Some caregivers could not or would not work the new schedules. They yelled that 'this schedule will ruin my family life,' and 'this project is certainly not good for caregivers, and probably not good for children either!'

"I asked them how many days a month they see the children on the current schedule. They calculated that it was about 10 days. So, I said, 'In 30 days, you see the children on only 10 days and not on 20 days of the month. How can you help children develop and have a relationship with you when you are there only one day out of three?' This seemed to give a few of them pause.

"Over the next few days we showed them videos from the institute in Europe in which caregivers are very child-focused, calm, slow-paced, sensitive, and responsive in their caregiving routines. They were very resistant. They dismissed those actions, saying, 'We do that, and we do it better. We did not learn anything from these videos.' They were defending their current behaviors and could not relate to, or understand, the benefits of the new approach. The discussions got quite loud and angry.

"So, we had them role-play several typical caretaking situations in which one person would be the caregiver and one the child, and the caregiver was instructed to behave as they usually do in the agency with the children. Then, another person was the caretaker and was instructed to behave in an engaged, emotionally positive, sensitive, and responsive way to another person who played the child. Some who played the engaged caregiver actually had trouble doing so. They could not go slowly; they were so used to moving rapidly. They could not wait for the child to respond or do something; they were so dominating and controlling. They can't wait for the children to be creative. They are impatient; they get anxious if they are not 'teaching the children.' One said, 'If I am waiting for the child to lead, what is my role? I am not doing my job!' They are so caregiver-directed, not child-directed.

"And caregivers were even opposed to interacting with children to build relationships. They did not seem to know how to interact unless they were making demands on the children. On the one hand, this is not surprising, because this is how they behave every day in the agency, but on the other hand, it is surprising, because I don't think they behave this way at home with their own children.

"Then, we asked those who role-played the children how they felt about the activity and about the caregiver. Of course, they felt much more positively about the engaged, warm, interactive caregiver. But many of the caregivers in the training did not see the point. 'How is this going to help the children?' they asked. But they were also fatalistic about the project, complaining that 'nothing in the agency depends on us; if the administration decides to do this, our opinion will not matter.'"

"I think the role playing was a good idea," commented Chris. "But why did you show videos from Europe?"

"Because these caregivers had been trained before," answered Sara, "and we needed to show them something different. I had visited that institution and was impressed with their caregiving."

"It would have been good for all of us to have discussed this before the training," said Chris. "Bob and I have visited there as well. They are very extreme and methodical about every interaction. There must be a hundred steps to supporting a child, and it's all written down. This was extreme to us; it must have been absurdly unreal to these caregivers. Also, showing them the results that illustrated that they did not do well in the original intervention would not likely motivate them to do better but to defend themselves and resist such changes."

"But it's done, and generally you did a good job. Did they ever come around?" asked Chris somewhat anxiously.

"Some did, partly at least, but some did not," answered Ellen. "It's going to take a good deal of constant supervision to get them to change their behavior with the children – it will take more money than we will have and more than three or even six or nine months."

Overtime payments. In a separate meeting, Alice, the Agency Director, told the group how many of each level of staff would be involved with the changes in the two groups of children. "In addition," she added, "there are several department heads in the agency, and this project touches all their responsibilities."

Alice and Ellen calculated how much money would be needed to pay all these people extra. It was more than the budget would allow. "Then, pay different amounts according to the person's level of responsibility," suggested Ellen.

"No, no," Charles asserted with some animation. "We are breaking our main principle. We should only pay people who do extra work, not because the project touches on people in their department. We should pay people at their current rate of pay only for the extra hours they work on this project. Yes, this will mean people are paid different amounts extra, and they will complain. Staff talk to each other and compare things like this. They will get upset. But at least you have a good rationale for who gets paid extra and how much for working on the project – period. And you do not need to pay me extra at all!"

"That's fine," countered Alice, "but all these people will have a role. They will change what they will be doing; they have responsibilities."

"Their jobs have changed a little in their particulars but not in their main purpose," observed Bob. "Caregivers will need to smile, listen, and engage with children, instead of just serving food and watching that they are safe, but their job is the same – they are still 'caring for the children,' and for the same number of hours per week. And a department head is not really doing anything more than before. These people should not get paid extra from the project, unless they work more hours."

"I agree," said Charles, "but special teachers will have to change entirely the way they operate, and they need to learn to do those things, work to do it, and it's hard for them. And some caregivers will have to learn how to position and handle children with disabilities."

"I think the project should pay for extra work and for training substantially new skills," added Chris.

"OK, I agree," said Alice, the Director. "But some staff have been doing for some time the new things the project wants staff to do, and they have not been paid extra."

"Maybe that is not really a new skill," wondered Sara. "Maybe no one gets paid for that activity unless it's extra hours."

"I think we are close to agreement on this," observed Ellen. "But we also need to think about the future. If we start paying people extra for things in this project, will we be able to continue to do that in the future? Have we just raised the salaries of caregivers in the agency indefinitely? Can we afford to do this for the remainder of this project and its extension, if there is one?"

We did the best we could under the financial and time circumstances and the personnel we worked with. It took several years of meetings and coaching, and, not surprisingly, the intervention was not as comprehensively or as effectively implemented as the original. Some caregivers became more engaged, but others did not. So, something was accomplished, but not to the extent that was desired.

Unexpected threats

Over the course of a major intervention project, a variety of things can happen that represent potential threats to the operation, outcome, and interpretation of the project. Some possible major issues are described in the chapters that follow, and we list a few minor glitches below to illustrate the kinds of things that can happen.

Renovations. During the course of an intervention, the local country's government provided funds to the agencies to renovate their facilities, including the sleeping areas, hallways, common rooms, and playgrounds. The agencies in our project were not renovated simultaneously, but rather one at a time. The work was quite disruptive, especially for groups of children whose rooms were being upgraded and painted. Further, the central administration sent fewer new children to the agencies when renovations were underway. This made it more difficult to implement some of the changes and even the caregiver–child interactions that characterized the experimental intervention, and it reduced the number of children we had to work with in all three agencies at different times.

Economic changes. Over the course of a project, economic conditions in the country of the project improved. This had several consequences. First, fewer children were sent to the agencies, so our population of children decreased in size. Second, the government did not reduce the caregiving staff at the agencies, so the number of children per caregiver – one of our major intervention conditions – was reduced in general across all the agencies in the study. Further, the agencies feared that if the number of children decreased sufficiently, the government would cut the number of caregivers. So, one of our agencies actually began to recruit children for their agency. These children tended to be healthier and better developed than the usual children.

Third, after a few years the government increased the salaries of caregivers by more than 70%. This wreaked havoc with our budget, which was fixed and could not easily absorb that magnitude of raises.

Changes in major administrators. In one project, the Head of the administrative unit that governed all the agencies in the city was replaced in the middle of the project. The new Head had no experience with the agencies or their services, and we had several meetings with him to acquaint him with our project and encourage his support of it. He had been an officer in the Navy for 32 years and enjoyed telling us how the Navy frequently outsmarted the US Navy. We joked with him about that, ingratiating his strategic cleverness. Our task was to have him support the project, so diplomacy won out.

Another potential problem was that two of the Directors of the agencies in our project were replaced or died during the middle of the project. This threatened the continued participation of those agencies or the nature of the intervention for them. Fortunately, the new Directors agreed to support the intervention that had been implemented in their agencies.

SPSS incompatibility. SPSS is a comprehensive database and statistical package of computer programs to which we converted the original data from EXCEL for analysis. Quite a bit of programming was necessary for data as voluminous and as complicated as one of the projects generated. When our international colleagues wanted to analyze data, they also needed to convert it to SPSS. SPSS was available in their country but under a different licensing agreement, and the program itself was slightly different. Trying to make the US and their versions compatible turned out to be a fairly difficult task. We sent our consultant to solve this problem, but he was not successful. Eventually, a technician from our partnering country accomplished the task.

City-wide training of caregivers. In another project, after the intervention had been in place for a few years and city administrators and directors of other agencies saw the improvement in caregiver and children's behavior, the city decided to offer training to the caregivers from all the agencies in the city. They even asked our international colleagues to teach social-emotional development and more engaged caregiving practices as part of this comprehensive training. This was a compliment to our project and good for children in the agencies, but it threatened the scientific quality of our project. Potentially, they were going to make all the agencies, including our comparison agencies, into treatment agencies, which might wipe out any differences we might see between the intervention and comparison agencies. However, most of the training was the usual health and safety information they always provided to caregivers. Further, the training was voluntary, and few caregivers actually attended. So, from a practical standpoint, it did not threaten the project's results, but the potential was there.

Other Issues

In a later project (thus we were calmer and had more experience), we lost all communication with our international partners, and the project did not have electricity for several days. There were public demonstrations that got out of control, and the police attempted to quell the rioters by shutting off electricity to the entire region. Fortunately, the blackout did not last too long and produced only a temporary shutdown of the project. Expect the unexpected, and accommodate as well as you can.

The Pandemic

Since 1918, there have been at least a half dozen epidemics and pandemics throughout the world with catastrophic effects on entire populations and countries. Most nations, including our own, were largely unprepared for the COVID-19 pandemic of 2019–2021. Concerned with our own country's response to the sudden disruption of life as we knew it, we neglected to put an ongoing international project into the perspective of what the pandemic was like in that country.

Specifically, several weeks into our US stay-at-home state policy to confront the virus, an email arrived from the project coordinator of our overseas project: "I need your advice; it looks like our country will be in quarantine for at least another month. Today I was informed that the environment in our primary agency will be modified to reduce the number of caregivers who care for the children in our project, and children will be moved into other groups! Further, we cannot get into the buildings to do the caregiver interviews. This will make it difficult, perhaps impossible, for us to assess the intervention's impact on the experimental group."

Both the nature of the intervention and how data were to be collected and on whom were threatened. The intervention attempted to make caregivers more consistent in the children's lives over time. Moving children to new groups and new caregivers was precisely what the intervention was trying to eliminate. Also, children would have new caregivers, not the same ones who made ratings on the children before the intervention started, so we would lose having the same caregivers rate the same children before and after the intervention. And assessors could not enter the institution to interview the caregivers and had to conduct the assessments by telephone or digitally.

Chris and Bob quickly debated the options, by phone, of course, since we were self-isolated. We would have to accept the online survey rather than the face-to-face interviews and direct observations that were planned. We agreed that the online survey done now would be preferable to waiting until the quarantine was lifted in that country. We did not know precisely when that could be accomplished, but it had to be done as soon as possible, before the group re-assignments contaminated the intervention. This was also going to take some creative data analysis to deal with the difficulty of tracing the youth and caregivers who were in our initial group and comparing them with children who were subject to the unexpected group changes. Sometimes, in the face of reality, one has to accept the best obtainable information when the best information is no longer achievable.

Our international partners agreed. They felt that it was more important to get results sooner rather than later, because they were working with political powers who wanted to understand the results and use them to justify critical changes in the service system countrywide. Of course, the results under these circumstances had to be positive, be reasonably credible, and support the proposed agenda of changes.

We recognized that to implement such sudden and seemingly extreme changes in methodology would require a great deal of trust in the long-distance relationship with our international colleague. We had to know that the data were collected faithfully and accurately in spite of the new methods. In this instance, we had a relationship that was already established and tested. We had confidence in our colleague. If that had not been the case or the stay-at-home orders had occurred a few weeks earlier and prevented the intervention from being implemented at all, these adjustments would not have been possible, and the project likely would have been scrapped. Of course, under other circumstances, perhaps we would have quickly taken advantage of our pre-intervention assessments and created a new project to study the effects of the pandemic itself.

We recognize that this event and course of action appear to conflict with the preferred procedures described in this book, such as being on-site or having a representative on-site, building in-country relationships in person, conducting multiple monitoring/supervisory visits, and visiting the country as often as possible to understand the culture firsthand. This project was deliberately designed to be conducted mostly by the international professionals with us as consultants. We had done much of the relationship building and training before the pandemic struck. Nevertheless, extreme times may require extreme and flexible responses to salvage a project and to do the best one can under the circumstances.

13 The Video Brouhaha

General Lessons Learned

The University of Pittsburgh Office of Child Development, which we co-directed, had a long history of funding for its domestic operations from a major foundation located in Pittsburgh. The foundation was quite enthused about one of our international intervention projects, but it was not accustomed to funding international activities. However, it thought this project would be groundbreaking from both a scientific and professional practice standpoint, so it decided to fund the making of a video describing the project and its effects on caregivers and children that could communicate the project's procedures and results to professional colleagues, funders, and policymakers.

We located a Pittsburgh Producer/Director and a Cameraman who had a superb relevant track record, including a production for the US TV series *NOVA* and videos on children. Moreover, their *NOVA* production focused on an issue in the same country as the project we wanted recorded, and the Producer/Director spoke some of the local language. Thus, they had the background and credits to produce a documentary video on our specific project. They were also very interested in the project from a humanitarian standpoint and were willing to produce the video for us at a very modest cost.

Discuss and continually inform your international colleagues about every aspect of a major communication project. We discussed this possibility with our international colleagues, who agreed that a video of the project would be useful. We briefed the videographers on the project and described the behaviors that existed previously and those that we tried to change in the intervention. The videographers rented gear, packed it up, and accompanied us to visit the project site and begin taping. The project was already in full swing, and the behaviors of the caregivers and children in the intervention agency were on full display, although the data were not completely available yet. They also taped in the comparison agency and interviewed us and our major international colleagues who spoke English. English was voiced over interviews conducted in the local language with some staff members.

There was no efficient way to collaboratively edit the emerging product long distance with our international colleagues. So, Chris and Bob worked closely with the Producer/Director in editing the footage into a self-contained video. We made suggestions on several rough cuts and edited the narrative. We tried to ensure that both of our major international English-speaking colleagues were fairly represented; that we and our international colleagues planned the project together; and that the portrayal of the agencies, caregiver behavior, and the project was accurate and representative.

We sent an early rough-cut version to our international colleagues, so they saw the general nature of the video. They only suggested some refinements in the English translations of their non-English-speaking colleagues. We showed a penultimate version to our project funders, who loved it, and we all teared up at the end with pride. Then, we took that draft video on our next trip to show our international colleagues; we were very proud of it and eager for them to see it.

But we were naïve about many aspects of this activity, and so were our international colleagues. Given the long-distance between us, we tried to keep our colleagues involved in the production as it progressed, but they objected vigorously when they saw the penultimate version. It conflicted with their professional and cultural values that even they had not anticipated until they saw it in its nearly final form. This experience does not need to be tied to a video per se; it could be over any substantial communication product or some other aspect of the project. It illustrates how individuals and cultures may differ in how and when they disagree. It is difficult for us to see how these disagreements could have been avoided except to have been clearer about the nature of the production from the beginning.

Illustration

The Preview

The anticipated preview showing of the video was a disaster, and all kinds of cultural and professional irritations exploded.

"We can't show this here," Irene objected with considerable tension in her voice as if holding back even greater emotion and anger. "It's a political video for general audiences; it's not for a scientific or professional audience. It would produce the wrong idea in our country."

"Irene is right," echoed Cynthia. "We cannot change our system of services without our own version of this video, not an American version."

"Can we have the video footage with no English on it?" suggested William in rapid succession. The tension and anger in our international colleagues were mounting.

"Why did we produce this video in the first place?" Irene asked rhetorically. "To convince US politicians to give money? See, then it is a political video."

"But what would be the purpose of creating your version?" Chris responded also rhetorically. "Cynthia just said it would be needed 'to change your system of services,' so your version also would be 'political.'"

"This video is not what we expected," observed William. "We have used scientific videos illustrating types of parent–child attachment that psychologists taped. They have been very helpful."

"Yes," agreed Cynthia. "We only need a few minutes to illustrate certain behaviors. I expected your video could be used in our country, but it cannot. The strongest part of it says that it's a sad state in our country and the United States can help to change it. It serves a US purpose – it could be shown on US TV. It cannot be shown in our country, and it is not necessary for my professional life."

"No, no," Irene countered raising her voice. "You can NOT show this on TV in the United States or here. It would hurt us; some people here could end the project. There is no purpose in evoking such emotions without information about the history, culture, people, economy. Our agencies served a valuable function after the war taking care of vulnerable children, and many of our citizens view them very positively. Americans do not understand this – don't show it to the public."

"The video is about the project, not our country," countered William. "It shows the changes that are possible – the other details will be forgotten."

"The current video is manipulative," Cynthia continued. "It manipulates the viewer's emotions; it makes them feel sad and negative. It's made by Americans according to American principles. It manipulates the viewer the way Americans advertise products. I only wanted a video for professional audiences that is a non-emotional report of the project."

Chris disagreed. "If I thought the video was 'manipulative,' I would not show it to anyone. Something is 'manipulative' if it tries to persuade people of something that is not true. Nothing in this video in inaccurate, out of balance, exaggerated, or gives the viewer the wrong impression. All of the people who have seen this did become emotional, but in a very positive sense. 'What a wonderful project,' they said. 'It brings tears of joy to my eyes.' Is that not fair and reasonable? Isn't that precisely the impression we wanted to convey about our project – that it is good science in the service of helping humanity? Seems to me that this was the purpose of the video we all agreed on."

"I agree," said William. "The video is emotional and not manipulative. It's the project that is emotional. Our own filmmakers would do the same."

"Yes, even worse," Cynthia quietly muttered in agreement.

"I think we need to recall our discussions at the start of the video project," Bob cautioned. "We all wanted to illustrate the positive changes that our project made in the agency and its services. A video would be so much more illustrative and persuasive than only data. We recognized at that time that to show improvements would also show that the current situation is not as good as it could be. Further, you worried that some of your politicians might see the video and make more trouble for the project than they already have.

You all were very nervous from the beginning about any of your colleagues or citizens seeing the video. So, you all told us that you did not want your own video, which is why we only made an English version.

"Further, this is not the first version of this video that we have shown you, and we sent the final narrative for your approval. There was an opportunity before for you to suggest changes, disagree with the style, and add cultural context. You made some changes in the translations of some non-English speakers, but you wrote to us that 'it is really good' and gave us permission to show it at a professional conference. You did not say anything then like you are saying now."

"There is another problem," Irene continued. "I see the copyright is owned by the University of Pittsburgh and all the video footage exists in Pittsburgh. The copyright and footage should be with the agency; our legislation says it should be here. Your videographers could take this and show it anywhere they wanted at any time. Also, parents of children who were in the agency temporarily did not give permission for their children to appear in the video. We could get put in jail for these violations."

"We understand your concern," assured Chris, "but it is not as bad as you think. You are correct that the copyright is with the University of Pittsburgh, because the grant to produce the video was given to the university, they own the video, and they arranged for the copyright. We would not know how to do that. The footage is with the videographers until they are finished, and then it will be at the Office of Child Development at the university. If you want copies of the original footage, we will have to see what it would cost to duplicate and ship it – there is quite a bit of it. The videographers have a contract with our university that says that they produced the video for the university and cannot show it on their own to anyone. The video itself states it cannot be shown without permission. We – the five of us – control who sees it. We agreed about that from the beginning. We were sensitive to children whose rights had not yet been relinquished to the agency and asked that they not be videotaped or not used in the final presentation."

"Who has a copy of this video?" Irene challenged.

"We gave a preliminary version to the project's funder and to the foundation that paid for it, and Cynthia wanted us to give a copy to a colleague in Israel who asked for it. We asked Cynthia if we had your permission to send it to Israel; we waited three weeks with no answer. We assumed we had your permission, so we sent it – maybe that was a mistake. We are required to give a copy of all products supported by a grant to the funder. All of these people were told they could not copy the video or show it to other audiences, and it says this in the video and on the disk."

Handwringing and Debriefing

Chris and Bob walked slowly back to their apartment at the end of this day, shocked and befuddled at what had just transpired. Once inside the apartment, we plunked down at the kitchen table and looked at each other in exhausted disbelief.

"What the heck happened today?" Bob declared. "I think we need a drink and to debrief.

"Here's to us," he toasted as we raised our glasses. "It's a cinch no one else is drinking to us."

"Why didn't they say anything before – now that the video is essentially finished?" Chris wondered shaking her head in disbelief.

"I don't know," Bob admitted. "But this is not the first time they have said nothing until everything is done and then disagreed. William said once that they don't criticize and make suggestions – it's a cultural thing. I recall that some other US colleagues who conducted international projects experienced a similar reticence by their international colleagues to criticize. Indeed, none of our colleagues criticized the video before this – they even said it was 'really good' and we could show it to an international professional audience. Now they strenuously object."

"Maybe so," agreed Bob. "But why are they upset now and not before? Is it really a cultural style to wait until everything is finished to disagree?"

"I don't know. I think Irene is upset partly over legal issues, and her concerns seem the most understandable. She is the one who could get in a great deal of trouble if there was a problem locally over the video, and word does get back here about things that go on in the United States. Because she does not speak English very well, she may not have known that we kept out of the video those children who were not yet legally relinquished to the agency. And because she does not speak English, Cynthia and William may never have shown her the rough cut or narrative."

"What about the 'legislation' that the agency must own the copyright, and the raw footage must be kept at the agency?"

"I'm not sure about the copyright business," admitted Chris. "It is hard to believe there would be legislation specifically dealing with the copyright to a video made in the agencies. There are likely laws about photographing the children, but Irene is authorized to give permission. You and I have recognized in the past that there are laws here that are not enforced, and there are 'laws' and 'policies' that don't actually exist – at least it seems that way."

"I agree," said Bob. "Remember when we asked whether families with children with disabilities received any benefit from the government? Irene said 'Yes, they get $100 per month.' Then, we asked – and we now have learned to ask this in these situations – whether the families actually get the $100? She said, 'No.' Why not? 'The government does not pay it.'"

"But in fairness," continued Chris, "they apparently do have policies about the anonymity and privacy of the children. So, she could be worried that some parent sees their child in the video and gets upset that the secret is out."

"Well, that could be a reason not to show the video in this country. But as a practical matter, what is the likelihood of this actually happening?"

"It probably doesn't matter," advised Chris. "If it's against the policies, it's against the policies. Look how upset some people here got when they thought – erroneously – names and tests scores were being sent to the United States and insisted that all data must reside in the agency. The video raises similar issues. We're going to have to deal in some way with this concern; we got all the appropriate permissions, but Irene may have reconsidered all this now that she has seen the video and thought more about it."

"But maybe there is more going on here," mused Bob. "They think it shows their country in a bad light, especially the way children are cared for in the agencies. We don't think much about criticizing the United States, but there is a long history of not doing that in many other countries.

"Let's step back a bit and try to take their perspective. On the one hand, I think they thought this was going to be simple video footage illustrating the caregiver–child interactions that the project promoted. What they saw earlier was basic edited footage, and they did review the narrative. But now the nearly finished video has music, titles, and credits, and it has a clearly flowing story to tell. It's a very polished production – it really does look like a TV documentary that is suitable for a lay audience as well as a professional one. It is not boring science; it is an interesting and heart-warming story – we've done a good thing. I don't think they expected it to look complete and sophisticated but rather simple raw footage of behaviors like the videos their colleagues make. Perhaps that's why they did not object until now. You did a good job countering the 'manipulative' argument and showing that the emotion created served the video's agreed-upon purpose, but next to their unexpected impression of the video it probably doesn't matter to them."

"I suspect you are correct about it having a popular-audience character that they did not expect," admitted Chris. "I think we need another drink...

"But I also think they perceive that the video portrays this as an American project that brought modern childcare practices to their 'backward' country. I know we tried to be fair about portraying them as equal partners and showing them an equal amount of time, saying that the project was jointly created, and so forth. But the video is in English, the caregivers were trained by an American, and you and I summarize the project's accomplishments at the end. It does look like an American project. I don't think there is any way around their perception, and I think we are going to face this issue over and over again in this project."

"I agree," Bob muttered. "It does look like an American project. So, what are we going to do?"

"Well, we agree that the video only can be shown to professional audiences, which includes funders, policymakers, and students, in the United States or at international professional conferences. Each user must sign a statement agreeing that they will not copy the video and will show it only for the specific purpose they requested. The video must be returned to us, and we will keep a record of who uses it. We agree it cannot be shown to the general public or on TV. Other uses must be approved in writing by all five of us. We may need a formal agreement with them to this effect."

"That's appropriate if not disappointing, but that's the easy part," observed Bob. "I don't think we want to give up the original footage, because it would mean we could not do anything else with this investment, but they could. I am not sure the university would relinquish it anyway. Duplicating it and sending them a copy could be quite expensive. Also, I am not sure they have the electronic equipment to edit it from scratch, but even if they did it would be very expensive and time consuming. But do they want their own version after all that harangue?"

"They seemed to equivocate about that" observed Chris. "They don't want THIS version, that's for sure. They don't want it in English, which is reasonable. They don't want some of the scenes of caregiver behavior that are common but are a bit aversive. They would minimize our presence and that of the United States, and they might want to add some footage of them and anything else they want."

"That would be OK it seems to me," agreed Bob. "Their best practical strategy would be to modify the existing video to fit their purposes."

"So, in addition to the usage arrangement," proposed Chris, "we say we will investigate the copyright issue with the university and explore the cost and feasibility of copying the raw footage."

After more discussion and emails following this visit, our international colleagues became more interested in having their own video tailored to their culture and preferred style. We all agreed that our expectations were not fully articulated and agreed upon from the outset, which then led to the sharply contrasting perceptions of the finished product. The specific proposals we made were accepted. A formal arrangement was drawn up between the five collaborators stipulating the usage principles, and a record was kept of its showings. Ultimately, the raw footage was not transferred or duplicated, and the original copyright was retained by the university, which gave blanket authority to the five of us for its use. A finished copy of the final video in a format suitable for international equipment was given to them.

It is difficult to imagine how this conflict could have been avoided, except by more discussion about the intent and form of the video at the beginning and more frequent sharing of rough cuts as the production process unfolded.

Eventually, our international colleagues did produce their own video using the original video as a basis. The had told us they intended to create their own video, but we were shown their video only once several years after its production. It is roughly in the same style as the original, but we do not have a copy of it, have no knowledge of the narrative, and have no record of its showings.

14 Publication Issues

General Lessons Learned

If the project will collect data of any kind, a first consideration is who owns the data and who controls its use. On the one hand, if the data are collected in another country, and especially if they are collected in an institution of that country, the country and the institution will likely have policies and laws governing who owns, has access to, and controls those data; what information cannot be communicated; and where such data may be stored and under what conditions. The European Union, for example, has fairly restrictive policies in this regard. It is best to check with your Institutional Review Board early on about such regulations.

In addition, the funding agency may have policies regarding data ownership, control, confidentiality, and sharing, and some of those regulations may conflict with those of the host country and institutions and perhaps with your own institution. The US National Institutes of Health and the National Science Foundation, for example, have such policies, especially with respect to sharing data that were collected with funds from the US government. Also, the US government and universities have become concerned about information and technology theft, and have established policies regarding what information, especially basic data, faculty can divulge or even carry with them on international travel to meetings.

Private enterprise will have analogous but different considerations regarding revenue from services or product sales, local taxes, import/export regulations, movement of funds across national borders, patents and copyrights, information and technology transfer, and so forth.

Your colleagues will also have ideas and opinions about their rights to the data. These need to be discussed early during the project planning phase.

The main advice is that you need to become aware of all these circumstances that potentially affect your project from the beginning. Waiting until the project is well underway risks a great deal of time passing, perhaps with severe consequences to you and the project.

Discuss publishing and authorship as soon as practical. A major end product of most academic and professional projects is a final report and mainly one or more publications. Projects conducted in some countries may be subject to local government review and even censorship, and some funders may also exercise review and control over publications. Most North American, Western European, and Australian/New Zealand professionals are likely to be totally unaccustomed to such oversight. Further, many universities have policies that insist on the right to freely publish results from research and professional projects as a condition for permitting their faculty to participate in the project. We suspect that universities have made some provisions for the lack of freedom of information exchange that exists in some countries, but you might consult with your university legal department if you are considering a project to be conducted in a country or funded by a local source that might exercise control over how the information to be gathered can be communicated and to whom.

Consider the form of reports and publications and their authorship. For academics, the journal in which the report is published and the sequence of the names of the authors hold weight in their professional evaluations. Some journals are more prestigious than others. American and European research journals are often viewed as the most desirable, certainly by US and European researchers and professionals as well as by universities in many foreign countries. But publishing in a journal of the country in which the project was conducted might be more readily seen by international colleagues and perhaps more likely to engender social, professional, and practice changes in that country. Generally, publishing the same article in two different journals using different languages is not common and can be a little complicated. For example, US journals may not permit an article published in another language to subsequently be published in English, although the reverse sequence may be allowed.

Finally, the report you prepare for a foreign funder or for your international colleagues who want to use it for practical or political purposes may need to be written in a slightly different style. One approach is to accompany a technical report with an "executive summary" which omits most technical detail and presents the basics of the project in common wording. That summary may be suitable for distribution without the full report to practitioners and policymakers. Alternatively, a single report might be prepared in common language with technical details presented in appendices.

Check the policies of academic rewards in the universities of your partners. Universities in other countries may have different policies and preferences regarding which journals and which kinds of publications (e.g., scientific article, book or monograph) are counted toward the academic standing of their faculty. Also, the nature of the authorship (e.g., authors named, a team authorship) and the sequence of names may be counted differently than in North American and Western European universities. For example, journals are now reviewed for quality, and those approved are listed in several services (e.g., SCOPUS). Publishing in a journal listed in one of these services may be a requirement for your colleagues. Again, if you intend to publish with international colleagues, it is best to find out these local university policies in advance.

More specifically, the sequence of authors on an academic publication has long been a major issue. Years ago, it was not uncommon in some disciplines for the Head of the Laboratory or even the Head of the Department to be listed as a first or last author on essentially every publication that came out of that unit regardless of whether that person made any contribution to the substance of the paper. More recently an international multidisciplinary task force issued guidelines ("Uniform Requirements for Manuscripts Submitted to Biomedical Journals") that have been adopted by more than 500 journals that describe who qualifies as an author and in what sequence authors should be listed. The guidelines expressly forbid such "courtesy authorships." The sequence of author names should follow their relative magnitude of contribution to the creative product. Also, as collaborative projects became more frequent, "team authorships" have also emerged in which a collective team name is the author, and members of the team are listed in a footnote with or without their individual contributions mentioned. However, such a team authorship may not be recognized by all international universities.

Try to respect local academic criteria. Although many of these practices are now commonplace in North America and Western Europe, universities in some other countries still follow older and sometimes different practices in awarding academic credit to faculty. Although some of these policies may seem anachronistic and sometimes silly to Westerners, they are real to academics in those countries. They represent issues to be understood and resolved. It is best to discuss publishing plans (e.g., journals) and who will be authors and in what sequence as early in the project as reasonable. However, sometimes author sequence cannot be easily determined in advance, because it will depend on who actually makes a creative contribution to the project. But some advance discussion might help create appropriate expectations.

Illustrations

We held some, but not all, of these recommended discussions regarding data ownership and authorship of publications periodically throughout most of our projects, and for a few projects, we even wrote out authorship policies that were occasionally revised. Nevertheless, these policies were a continuing source of conflict in some situations as described below.

The Data

You may find that your international colleagues are well aware of laws and policies regarding what information can and cannot be shared with you or other people and released by agencies in their country. Those policies may or may not mesh well with US privacy considerations and your need for certain pieces of information. For example, you may not need the names of children or service personnel, so they may be assigned subject numbers and your international colleagues keep the key. Further, raw data records may remain in the agencies or wherever they are collected, and only numerical data taken from those records entered into the database that is shared with you. Other information may be heavily restricted. Nevertheless, these practices and other safeguards may not prevent certain people from making threatening and false accusations in an effort to shut the project down.

Chances are that data will be collected and entered into EXCEL files or some other electronic platform. Then, the data may be transformed into a database management program, such as SPSS, that permits statistical analyses. Decide whether your international colleagues will make this transformation, or you will do it, and make sure that your database management program is compatible with your international colleagues' program, even if they have the same name. It is likely that your international colleagues will want a copy of the database management program and will want to analyze some or all of the data. Hold several discussions about data analyses, and try to get an idea of how skilled and experienced your international colleagues are at the kind of data analyses that are likely to be needed for your project. They may be quite knowledgeable, but not very experienced at analyzing large datasets. Also, recognize that familiarity with certain statistical techniques is only the first step in analyzing data to answer research questions and pass journal reviewer's criteria.

Authorship Dilemmas

Sooner or later - preferably sooner - you will need to have one or more discussions about communications from your project, which are often in the form of publications, either scholarly articles for journals or general audience reports of the project or both. Here again, you will need to judge your international colleague's experience in writing such papers, especially if they are to be in English, which may be a second language for them. Your collaborative project may provide a good opportunity for you to mentor your colleagues on writing and publishing a scholarly paper. But be mindful that your international colleagues may recognize that you have much more experience in writing such papers and defer to you even though they want to write the paper themselves and may feel it is their prerogative because they implemented the project in their country. But they may be reticent to suggest that they write the paper. You may need to be sensitive to this situation and ask them directly what role they want to play in publishing.

In addition to who actually writes the paper or report, you will need to decide on its authors – who is an author, and in what sequence are they listed?

Generally, to be an author, a person must have made some contribution to the creative product of the paper, and people are listed in order of the magnitude of their contribution. These principles are a rough guide at best, and they don't solve all the problems, because different "contributions" are not comparable like apples and oranges, and there is no metric for measuring them. Some diplomacy may be required to determine the authors and their sequence.

Consider the following possible events that illustrate some of these issues. Suppose there are five major contributors to a project, two American and three international partners. The international colleagues insist that all five partners be listed by name on each major publication reporting on the project's main findings. This is certainly the conventional procedure, but it leaves open the crucial issue of the sequence of authors named, especially who will be the first author. Our international colleagues suggested that we alternate first authorship – an international colleague would be first author on the first paper, an American would be first author on the second paper, an international partner would be first author on the next paper, and so on.

This strategy seemed to satisfy our international colleagues, but it was better in theory than in practice. It ignored the nature of the article, and it potentially violated the international agreement on the sequence of authors. What if the paper was mostly the work of the international group but it was an American's turn to be first author?

The issue came to a head when the group was asked to contribute an article to a special issue of a journal. The article was to describe the nature of the children, caregivers, and characteristics of the agencies we were working in. It was essentially much of the baseline data we had collected before implementing the intervention, so it was a main report of the project.

Chris and Bob felt that a team author would be best – "The International Services Team," for example. This avoids conflicts over who is to be the first author. Team authorships were becoming more common in the United States, and some teams have become quite well-known and prestigious. Individual faculty who are part of a research team would list each team-authored paper on their professional resume and describe their role in producing each paper on their annual evaluation report.

Our international colleagues objected. Team authorships were essentially unknown in their country. They would not get any credit from their university for a team-authored paper. They had to have their names listed as authors.

Chris replied, "We understand that a team author is unusual, but we all must follow the international guidelines for who is an author and who is the first author. There is no question that all five of us should be authors – we have all creatively and substantively contributed to the information in the paper. Further, following the guidelines of who has contributed the most to this paper, there is no question that Bob should be the first author."

Our colleagues would not agree to that.

"Then, accept having a team author," concluded Chris. "How about this: In team-authored papers, the members of the team are listed in a footnote on the first page of the paper. What if we list the team members alphabetically first by country and then individuals alphabetically within country. Then, all of you will be listed first, and Bob and I will be last in the footnote."

We can't say our international colleagues were happy with this "compromise solution," but they accepted it for this paper.

This experience required that we consider authorship for all kinds of future papers and communications from the project. So, Bob and Chris drafted a seven-page document that described authorship policies for essentially all the issues and options. It defined five types of communications and presentations that group members and others (e.g., students) might engage in (e.g., main report of the project, secondary reports, colloquium presentations), criteria for authorship, authorship options for each type of presentation (e.g., individual named authors, a team author), the advantages and disadvantages of each alternative, and procedures the group should follow for each type of presentation. It recommended a team authorship of main reports from the project and specified that authors be named for all other kinds of papers and presentations.

This set of policies stood – until the next main report of the project was written a few years later.

It was clear that reporting the results of this project would require a book-length publication or numerous individual articles. Bob argued that it was wasteful and inefficient to publish separate articles, because the intervention, sample of participants, and procedures would have to be repeated in each article or summarized with a reference to another article describing all the details. Also, one loses the flow and relations between different aspects of the project and its results. He suggested the results be published as a book or monograph.

The international team members again favored individual articles with named authors, and they asked if separate chapters of the book or monograph could be authored by individuals. Bob reported that the monograph they intended to use had an informal policy of not publishing a set of separately authored chapters, but he could ask the editor if the group wanted this. However, this strategy would not solve the issue of who is first author on each chapter. A few chapters would be obviously assigned to the international colleagues, others to the Americans, but most would not be obviously one or the other and we were back to the old conundrum of who is legitimately the first author of these chapters. This is why a team author for the entire monograph is best, Bob and Chris argued.

The international colleagues reluctantly acquiesced. But when the monograph was published, they were upset anew. The cover of the monograph gave the title and the author as the team author they had agreed on, but then below that it named the editor of the monograph series.

"It looks like he wrote the monograph not us," our colleagues complained. "And the list of members of the team is on a page way in the back of the book, not up front where authors usually appear. This is totally unsatisfactory from our standpoint."

"Well, we have to agree with you," said Bob. "It is not what we wanted and expected either. But the undesirable characteristics of the presentation of the authors applies to us as well as to you."

Another issue arose simultaneously. Our colleagues wanted the report to be published in their language in addition to an English presentation, which they understood that we and our funder would require. We all agreed that would be good. Bob volunteered to ask what the policy was on having both an English and another-language publication.

The answer was a bit ethnocentric: If the manuscript is published in another language, it cannot be simultaneously or subsequently published in English in the publication we were considering. However, once it is published in English, it can be published in another language as long as the English-language publication is cited as the original.

Bob was upset and embarrassed by that policy, and he felt that it was not welcoming to professionals from other countries. Nevertheless, we all agreed that our international colleagues should explore the options for a local-language publication and try to raise the money, if needed. Eventually, they were successful. A major stumbling block throughout all this was the policies the foreign university used to give its faculty academic credit for publishing scholarly articles. There did not seem to be a policy in this regard during the early years of our project, except that articles authored by the faculty member were recognized whereas team authorships were unknown and not recognized. It was small wonder our colleagues resisted a team authorship.

Long after the monograph was published, their university adopted a new policy in this regard. Indeed, they now required publications for faculty advancement. But they gave credit only for journal articles and only those published in journals recognized by the international listing services. The sequence of the authors names on the article was not relevant, only the number of such articles. A computer reads the footnotes, and if members of a team author are named in the footnote, they will get credit, so a team authorship could be used with an appropriate footnote. But books and monographs do not count.

"Well, that is unique," opined Bob. "It seems to ignore the very long tradition of scholars publishing books, which are the major academic products in some disciplines. Oh, well, we don't intend to publish another monograph or book, and we will probably not need to use the team authorship again."

But the publishing issue was not totally resolved. An article authored by the group with individuals named was submitted to a relatively new journal published by a major American professional society which sponsored numerous other well-respected journals. Our local colleagues complained that this journal was not on the lists of approved journals accepted by their university. Could we submit the article to another journal?

Bob wrote to the editor inquiring why the journal was not listed and suggested how important it would be for a journal specializing in international articles to be listed. The editor agreed and said that it was a new journal and it was in the process of applying to be listed. This satisfied our colleagues and their university.

Report vs. Publication

A totally different issue arose in another project in another country. The funder of our work there was in turn funded by an agency of another government. Near the end of their grant period, the agency who had hired us was required to submit a final report to their funder. Their report was due quite soon, and they asked us to complete the data analyses as soon as possible and write a report that they could submit to their funder.

We dropped everything, and several of us hurried to analyze the data. We got them a report on time. Later, when we were less rushed, we prepared an article for publication, which our funder was eager to have done. But the demands of journal articles are a bit more stringent in some respects than reports to funders, so we made some additional decisions about including a few more children in our analyses who had been omitted from the previous report. This changed a few numbers very slightly, but not the general results of the project.

When the article came out and we sent it to the funder, its representative was furious that a few numbers were slightly different in the publication than in the report they had submitted to their funder. How was that to be explained to their funder? Had they sent an inaccurate report to that funder? We tried to explain the nature of the differences, why we made those changes, and mainly that the results and conclusions were the same. This explanation was not acceptable to this representative, but nothing ever happened as a result of this situation. Perhaps their funder never saw the publication and/or did not notice these inconsistencies. But the situation tarnished our relationship with that funder for a long time.

"You Can't Publish This Report"

Still a different publication issue came up in yet another country. After the evaluation of the funder's intervention was completed and an article for publication was drafted, the foundation decided that it did not want the evaluation published. There were certain political issues rumbling in that country that potentially would influence the foundation's status and ability to operate there. Indeed, we attended a conference of all the people affiliated with the foundation and those representing agencies in which the foundation's programs operated. The President of the foundation gave an address at the conference in which she announced quite unexpectedly that the program as it then operated would be discontinued in a few years and that the agencies themselves would have to take it over.

Although we do not know, it is possible that political events stimulated this decision. The foundation had told us that they would ask the local government for permission to publish. Perhaps they decided this was not the time to ask the government to publish a report on a project that they were going to phase out. Also, while the evaluation showed that their program generally did improve children's development, the pre-intervention assessments showed how poorly the children in these agencies were developing without this program. Perhaps this was not a good time to publicize internationally that this country's services were not doing a good job. Ultimately, we do not know why they decided against publishing the paper.

This experience raises an important issue for US universities conducting projects in countries that do not share their value for the free exchange of research information. As noted above, most American universities have policies that essentially state that the university and its faculty have the right to publish the findings of their research, and without that right, they cannot conduct the project. There are a variety of good reasons for such a policy ranging from academic freedom to the tax status of universities. But in some countries in which the government exerts greater control over much of what transpires in society, the government may have the right to control what happens in the project and what is communicated about institutions that it governs. Similarly, funders may have the same right of control over publication. On the one hand, most university contracts stipulate that the faculty member has the right to publish the results of the project, and without that clause they will not agree to do the project. On the other hand, researchers and universities, which otherwise are often eager to conduct international collaborative projects, need to recognize in some way that there can be funder and political influence exercised when working in some countries that would not exist or be tolerated at home.

15　Project Death Threat

General Lessons Learned

Obtain the approval of ALL the relevant administrators and political figures before the project starts. A common reason for conducting a project in another country is to implement a program or intervention in another cultural environment, such as in another community, organization, or population. In contrast to projects conducted in a laboratory in a university, community-based projects often require the cooperation of several individuals and organizations – community agencies, administrative and political figures, additional ethics committees, community or private funders, and others – who have the power to approve or deny the right to conduct the project in the first place and who could potentially close it down in the middle of its implementation.

The reasons for such actions can be very legitimate, such as new local regulations; they are part of the collaborative process of conducting projects in community settings. But such threats also can occur for less than legitimate reasons – for example, politics, disciplinary prejudices, personal self-interest, power and control, financial gain, and jealousy. These factors can be vexing if the cultural environment allows certain actions that are less common in your home country.

The fact that a project requires the initial and continuing support of certain groups or individuals means that approval, often in writing, is needed before the project is initiated. This is obvious in the case of formal ethics reviews, approvals by administrative governing bodies, and the like. But there also may be individuals whose positions are related to the project but who have no formal or informal involvement in the project. These people need to be informed of the project in advance, but because they are not directly involved in its implementation, they usually only need to know it is going to be implemented and give it their informal "blessing."

Keep relevant stakeholders continuously informed throughout the project. But even when all these approvals have been obtained and the project is fully operating for some time, one or more of the groups and individuals who initially gave their approval may become less supportive or may be replaced with new personnel. It might be helpful to keep these groups and people involved in the project as it progresses, with periodic updates, progress report meetings, even site visits.

But this can be a two-edged sword. On the one hand, they will not be surprised at some point well into the project, and they may feel they have some participation in the project; on the other hand, you are inviting them to comment and make suggestions which may be difficult to ignore, and they may not have the time or interest in reading updates or attending meetings unless they do have some potential influence. Perhaps written progress reports or occasional private meetings are compromises; benign neglect may work out all right – or maybe not.

Illustration

What follows is a scenario that could happen. It represents an attempt to shut down one of our intervention projects after it had been operating for nearly two years. Suppose the project was already showing clear signs of improving caregiver–child interactions and children's development that visitors to the agency could readily observe before the data were available to document these benefits. The project was becoming more influential than many people initially expected, and it could no longer be treated with benign neglect. The seeds of this threat began early and were discussed by the leadership team at one of its periodic meetings. Eventually the threat took a very concrete form.

"Our project has been 'discovered' so to speak," Paul, the Director of our main intervention agency, reported grimly. "The two administrators who are in charge of medical and educational services in all the agencies across the city are now paying attention."

"Yes, we remember meeting them," Chris replied. "We saw them early on and then when we sought their permission to do the project. We described what we wanted to do and in which agencies, and they approved. Actually, they seemed rather disinterested in the whole thing. It was as if they thought this was a trivial project that wouldn't change very much, so 'go ahead.'"

"Well, they are not disinterested any more, especially one of them," asserted Paul. "She hears from one of the other Agency Directors about the changes in my agency. She also visits other agencies, and she has heard from some of their personnel, especially ones in the other agencies in our project. They are very highly regarded and powerful in town, and some like our project and some do not. Also, numerous professionals from around the city have visited our facility and have seen the difference in the way our caregivers and children behave."

"We knew early on," mused Bob, "that whenever you try to improve something, like an agency, you imply that the status quo is not very good, or at least not as good as it could be, and the status quo is their responsibility. So perhaps the visible success of our intervention insults and threatens them."

"That may be," continued Paul, "but it is more than that. There are rumors that at least some administrators want to close down the unit in which these critics work, and then they both would be out of a job. They notice that the Directors of two of our project's agencies are both in their 60s and may retire soon, and they would like to be made Directors of those agencies if their unit were closed. I don't think that would be good for the project."

"Absolutely," added Harriet. "Remember that one of these administrators is in charge of medical services in all of the city's agencies, but our project is a behavioral intervention. That may be why she did not pay much attention to it at the beginning, because the agencies are operated by the local Ministry of Health and they don't pay much attention to behavior, except education. But now that people see how much better the environment is in our project, she's against it."

"She sure is," continued Paul with some anxiety. "There are rumors that she wrote letters to the higher administration suggesting that our intervention agency be made into a facility exclusively for children with AIDS. Although some of our caregivers would not want to work in such a facility, the salary for caregivers in a specialized agency would be 80% higher than they currently make. This would be attractive to some. But if this were to happen, all our current children would be transferred elsewhere in the city, and the project would be over.

"But it's not a very practical idea in any case. We already have a home for children with AIDS in this area; another one should be located elsewhere. Also, rooms in our facility are not big enough for children birth to 18 years, the configuration of rooms is not medically suitable for handling AIDS children, and our outdoor space is not big enough. So, it seems to me that in reality it's an obvious attempt to sabotage our project."

"This change would have to be approved by several high-level committees in the city," advised Harriet. "We met with these groups twice at the beginning of our project, the Head of the City Committee has visited our agency, and he says he supports the project. He suggested that we get letters of support from other officials, which we did, and we gave a presentation to the Board of Pediatricians and got a letter of support from the Head of that board."

"Yes, but there is more," Paul added impatiently. "One of these administrators who is complaining about the project has declared that she should be the 'Administrative Director' of our project."

"I think she is upset that she is not being paid by our project," Harriet speculated. Apparently, the Director of one of our other agencies suggested to one of the administrators that she should be paid by the project, and that message probably got communicated to the other complaining administrator as well. They probably know that the Directors of our agencies get a stipend for implementing and managing our project and the assessments in their facilities, and they know Paul gets paid by the project. Why not them?"

"I told them," Paul responded with increasing agitation, "that the project only pays people if they provide a service to the project. We decided at the beginning that although it is not uncommon in our country for payments to be made to administrators and politicians, we would not do that unless they actually worked in the project."

"And our funders would not allow it either," declared Chris.

"One of these individuals is a little more favorably disposed to our project than the other," observed Paul. "Her responsibility is the education of the children, which is behavioral. She showed interest in the data forms being used in one of our agencies, but I told the Directors they are not to encourage her probing into the project. On the one hand, she is telling other Agency Directors that caregivers need to get on the floor with children and engage children face-to-face, like we do in the project, but generally her ideas about how children should live in the agencies are very different from those of the project. She told a meeting of Directors that our intervention agency is organizing a home-like atmosphere, but she argues that agencies need to teach children to be prepared for life in the next public service, and a home-like atmosphere will not do that. I've taken both of them to conferences and other meetings about our project, but I don't have any hope of changing their minds.

"Both of them are also upset that we hired people to be primary care-givers who were not certified assistant teachers, so they claim our project is not educating children. I told them that such people received the same orientation training and that there were plenty of assistant teachers – in fact our agency has more assistant teachers per child than any other in the city."

Bob attempted a summary. "There may be three underlying themes here. First, these critics are embarrassed and perhaps threatened because they are in charge of services for children in the agencies, the project is doing a better job of caring for the children than they have done in the other agencies, and people recognize that. Second, because they are in charge of services for children and the project is providing services for children, they think they should be in charge of the project. Although initially they did not think the project was important, now that other people perceive that it is being successful and want to do similar things, they can no longer ignore it and want to be in charge of it. Third, they may want money from the project."

"I agree," said Harriet, "they may want to be paid. But they are in charge of medical care and education, and the project is not focused on these two aspects of care. They don't think the project will benefit children's development in those areas."

"They might be surprised some day," mumbled Bob to no one in particular.

"Also," added Paul, "they have been told – probably by a pediatrician in our agency with whom one of them communicates frequently – that our children come down with more infectious diseases, sore throats, coughs, vomiting, colds, and pneumonia, probably because they have more social contact with other children and caregivers. They do have more of these illnesses now than they used to have."

"A medically oriented administrator could use this against the project, saying it 'makes children sick,'" asserted Chris.

"We actually expected something like this to happen," Bob recalled, "and maybe we should have warned them about this. Studies of young children who were put into contact with one another in experimental childcare centers in the United States did come down with more such illnesses because of their proximity to one another. But the pediatricians involved in those studies actually concluded that this was good for children in the long run, because they would develop immunities to such conditions, and they were actually healthier later in life. I'll get the references for those studies."

"How should we handle this?" asked Chris. "I do not think Bob and I should be directly involved. If control is an issue for them, then having the US Team present would only add to the impression that other people are controlling things, especially Americans."

"Yes," confirmed Harriet. "I think our group should cooperate with them, not be confrontational, and give them information about the project and how it is being implemented. Unfortunately, one is very upset that Paul is the 'Administrative Director,' and she will not meet with Paul or return phone calls from him. However, she seems to get along with me and even confides in me, so there is a possible road to cooperation.

"We also need to cooperate with the city committee that oversees all the services in town. They have heard that we have presented the project to international conferences and showed videos of children being fed in the traditional manner, which is hard for people to watch, and it embarrasses them."

"I have also been told," Paul adds, "that one of them showed the administrators an article she said we wrote in a pediatric journal that was critical of the traditional services and that those criticisms were not true. She was angry about this article and that we did not tell her about this. She is also upset that a major component of our intervention is to mix children of different ages and those with disabilities in the groups, and she claims the Head Psychiatrist has stated that this is bad for children with disabilities."

"Research shows that also is not true," reminded Chris.

"Somehow we need to make them feel more a part of the project and help them take credit for it," advised Harriet. "Maybe we could get the data from them on the numbers of children in our project agencies that we periodically have to report to our funder. Their Center keeps track of that. It's not much, but it is some involvement."

"We need to plan a renewal project and grant application soon," observed Bob, "and we could have them contribute certain services that are necessary for that new project, if they will provide them, and give them a fancy-sounding title of some sort. This provides them with a stipend, but it requires some work and does not represent any control over the project, if that's what they want."

"We need to arrange a meeting with them and the city administrators. Let us outline the agenda," proposed Harriet. "We need to gently remind them that we discussed this project with both of them before it ever started, and they gave their approval."

Paul raised several accusations that he believes they have already made to the city administrators about the project that we should be prepared to refute. "For example, they have reportedly said that the project is not very professional; the assessments are subjective, unreliable, not valid, and not culturally appropriate for our children; and we will lie to make the results look good."

Harriet countered. "The project and all assessments were reviewed and approved by many scientists and passed by scientific and ethical review committees at the University of Pittsburgh and at our university. Most of our assessments are commonly used and valid, and they are administered by trained professionals whose reliability has been documented. Most assessments are more reliable than some common medical measures, such as Apgar scores. For us to lie about results would end our careers – it won't happen."

"They have also claimed," Paul continued, "that we have no right to send data on our children out of the agencies to the United States – it violates confidentiality."

Harriet again rebutted. "All names of children are removed from the database, raw data are kept in the agency, the children's current IDs are not passport ID numbers, and the key linking those IDs is locked in the agency. So, no identifiable information leaves the agency and certainly does not go to the United States. These procedures are reviewed each year by the Dean at our university and reported to the US funder."

"They will want to be informed of every aspect of the project," Paul warned.

"Well," advised Harriet, "they can visit the agencies in our project and attend our meetings and trainings, but the data and the budgets are confidential. They can talk to Chris about the budget, but there is no money in the agencies, and all expenditures must meet regulations set up by the US funder and the University of Pittsburgh."

"Then, she'll say 'so what? When the project is over what do we have? Who cares?'" Paul shrugged.

"The project is likely to show," Bob answered, "that improved early warm, sensitive, responsive caregiving and relationships in the services produce better social, emotional, and mental development in the children, even better height and weight. If this is true, such interventions should be implemented more widely in the city, country, and elsewhere."

"I'll set up a meeting with the administrators and city committee," Harriet volunteered. A few days later, the Head of the City Committee decided against such a meeting. Instead, he appointed a subcommittee of professionals who will review the project and report to him and the entire committee. Chris and Bob returned to the United States, but communications with the international team continued over telephone and email.

Several days later, Harriet called Bob and Chris and reported, "There was a meeting of the committee in the district that includes one of our project's agencies. Each agency is in a different district of the city with its own administration. The local district committee declared that the project must stop in their agency! They cited several reasons. First, they did not have enough documents, such as the contracts between the universities. Second, the data forms contain a child's diagnosis, date, and where the child goes after leaving the agency. The law states that such information may not leave the agency. Third, because the project pays people, it is considered 'commercial' and therefore must pay rent for the rooms it uses at the agency facilities."

"We only attended this meeting," Harriet continued. "They did not seem to be ready to listen to us or hear that some of these accusations are not true. It was as if someone had told them these things about the project; I doubt any of them had any first-hand knowledge about it."

"Oh no," Chris said in exasperation. "Then what happened?"

"Recall that the city administrator decided to form a special ad-hoc committee to study the project and present to the city committee the legal status of the project," replied Harriet. "That committee consists of a leading pediatrician, psychiatrist, and the two accusing administrators."

"Oh, wonderful," moaned Bob sarcastically. "Nothing like having the same people be prosecutor and jury, but I guess only we worry about such things. I suppose the two accusers do have administrative authority over and presumably knowledge of these activities in the agencies."

"We prepared all the documents for this meeting," said Harriet. "For two hours they said what they wanted to say. They said the City Committee on Health should approve the project because it is operating in the agencies which it oversees. The agreements and documents we have are not sufficient. They wanted copies of the agreements with the university and our agency."

"We can FAX those to you right away," offered Chris.

Chris and Bob immediately called a professional consultant who had extensive knowledge of social and political operations in this country. After being apprised of the situation, he advised:

"Well, I am surprised something like this did not happen sooner. My advice is to step back a little. Get this city committee's issues in writing; otherwise, it will be a moving target. Get the local university involved. They have clout, and they have people who have dealt with this kind of thing."

Bob and Chris had read about other situations in this country in which the government police, after accusations were made by jealous colleagues, stormed a project and confiscated all the data and computers. So, we urged our international colleagues, as best they could, to 1) backup all the data all the time, 2) try to enter as much of the collected data into the database as soon as possible, 3) make at least three copies of the database and keep it updated as well as possible, 4) store one copy of the database in your home and one in the home of someone not related to the project, 5) send us a copy of the database more frequently than usual, and 6) move all the raw data without identifiers to a safe place outside the agency until it can be entered (we do not want a police raid to take it, which has been known to happen). The entire project shut down for a month while they executed these protection measures.

Eventually, our international colleagues saved the project. They provided enough correct information and documents, and city committee members visited our intervention agency to see the project's effects first-hand. As a result, the committee approved the project and even asked to be kept informed and given a final report to determine if such procedures should be implemented elsewhere in the city and the country. The accusations from the lower administrators ceased, and eventually they even supported the project on occasion. No one was paid any "fee" or hired.

Never underestimate what can happen in projects like these.

16 Relationships and Internal Threats

General Lessons Learned

Strive for an "equal" partnership. This has been a continuing theme in this book. It is likely that many projects will be conducted as partnerships. Ideally, a true partnership consists of individuals who share a common goal for the project, each of whom has something necessary to contribute. Each partner must be respected for their necessary and unique contribution. There must be mutual trust among partners, and each is treated as an equal partner.

In practice, however, equality is rarely totally achieved, whether the project is a university–community international collaboration or one being conducted by private enterprise. There can be several reasons for this, including who receives and controls the funding for the project, differences in experience and status within a company or organization, contrasting skills and experience, different roles and responsibilities within the project, cultural differences in professional practices and behavior, and individual differences in personal behavioral style. These are commonly experienced issues, and conflicts pertaining to them can arise and even destroy a project.

It takes a leader with a special temperament and style to hold a collaboration together. That person must balance understanding, respect, and trust for each member of the group and their perspectives on the one hand, with an ability to get the group to make decisions in the face of disagreements and move the project along on a reasonable time schedule on the other. Depending on the size of the group of partners, no leader may be formally designated, and the group proceeds more-or-less by pure democracy. But eventually disagreements, even conflicts, are likely to emerge that demand decisions in the face of disagreements. A designated leader might help smooth this process, or sometimes someone in the group assumes some responsibility for solving a problem and coming to a conclusion. Ideally, you should pick your partners carefully, not only for their skills but also for their temperament and ability to work with people who have different perspectives. But chances are you will not be able to pick all or even any of your partners – they will be dictated by their positions, responsibilities, skills, and other factors.

It's not just a leader who needs these attributes; each partner needs them, too. All the members need to be able to listen to one another, try to understand and respect different opinions, be willing to compromise and come to a conclusion in the face of conflicts, and respect the other partners even after a conflict is resolved. Ultimately, there needs to be mutual trust among all the partners.

Strive for frequent, frank, complete communication. One key element in this process is relatively frank communication, especially when the opinions of partners conflict with each other. Many people, perhaps especially many Americans, are accustomed to open communication and readily state their opinions, and some are also accustomed to being overruled and getting on with the process. These characteristics are not as common in some cultures. There may be a deference to certain people – the leader, the elderly, the more experienced or higher status member. Opinions may need to be asked for in the group or later in a personal conversation. Some people may not have much experience disagreeing, and disagreements may be voiced in unfamiliar ways, perhaps aggressively and in an ad homonym manner or meekly and incompletely. It helps for partners to be sensitive to these differences as early in the partnership process as possible; otherwise, the group can get far down the road on a project component before disagreements come to light.

Take steps at the beginning to promote good social relationships among partners. Groups often take some time for the members to get to know and relate comfortably with each other on both a personal and professional basis. It may help for the group to engage in social activities together, especially at the beginning of the project. Engage in shared fact-finding early on, go to cultural events or sightseeing together, share meals with one another, and even hold after-hours parties together. This is a continuing process over the course of the project, because decisions and potential conflicts can erupt at any time, even years into the project.

How these people get along, understand and trust each other, communicate, tolerate disagreements, and work to resolve conflicts will be crucial to the success of the project.

Illustration

Recall from Chapter 1 that the US National Committee for Psychology identified numerous areas in which members of an international collaboration may differ in opinion and practice. These areas included differences that arise because of who controls the funding for the project and differences in the professional backgrounds and experience of team members. Also, conflicts can emerge over different views about who controls and owns the data, authorship issues, roles of each member of the collaboration, and communication styles and practices.

We have tried to create an example in which conflicts surrounding these and other issues arise. Although in reality one or another issue is likely to come to the fore at a time, for convenience in our illustration below, we have merged them all into a single episode. The illustration assumes that Chris and Bob plus three international colleagues have collaboratively created an intervention in a local agency designed to improve the care and services provided to young children and families. Assume that the project has been operating for a couple of years before the episode described below occurs.

Make a few additional assumptions that help to set the stage of this narrative. The project was funded by an American granting agency with Chris and Bob as the Principal Investigators. This is a common situation, but it means that Chris and Bob have the ultimate responsibility for the scientific and financial performance of the project. That is a big inequality between partners, even if that responsibility is never used overtly to leverage influence over the conduct of the project.

Another common occurrence is that there are substantial differences in skills and experience between the partners. On the one hand, each of those different skills is necessary to the collaboration, but on the other hand, they could be a source of friction between some members. For example, in the scenario below Chris had decades of experience planning, creating, implementing, and managing collaborative service interventions with community groups. Bob had 35 years of empirical research and data analysis experience with infants and young children, and both had held numerous grants, knew the required administrative and fiscal procedures, and had published literally hundreds of scientific and professional books and papers.

Suppose the international team consists of George, a Services Administrator who directed the local service agency in which the intervention was being implemented and knew how to manage it within the regulations prescribed by the government, William has experience with children and staff who work with them, and Jane is an accomplished scholar and psychologist. Both William and Jane are affiliated with a local university. Clearly, each brings necessary skills to the collaboration, and while there is enough overlap among them to produce understanding and mutual support, that overlap also contains differences of opinions that can lead to internal frictions.

Further, assume, as is common, there were cultural differences in professional education, practices, and behavioral styles. These differences may be part of the reason for the collaboration. For example, suppose that Jane and William were trained as psychologists and educators, but their professional training plus the academic procedures and rewards at their university are quite different from those of Bob and Chris. Jane and William have had some experience working in Western universities and services, so they have some knowledge of Western academic practices. Chris and Bob, however, are naïve regarding services and university policies in the country in which the project was being conducted. These differences can produce disagreements on the nature of each person's role in the collaboration, the intervention procedures, measurements, data analyses, and authorship, for example.

Also, at the risk of overgeneralizing, assume that Chris and Bob, as Americans, tend to be behaviorally quite open, direct, and politically naïve, whereas our international collaborators are much more reserved, are less open and forthright, and operate as independently as they can. Of course, these dispositions can be much stronger in some than in other individuals, but they can lead to communication problems and differences in expectations about certain things.

The constructed narrative presented below initially erupts openly one day during a meeting of the collaborative team, but it stretches over several days and then months thereafter. It illustrates how these factors can contribute to irritations, frustrations, conflict, and anger among some participants and shock and hurt feelings among others. Collectively these emotions can threaten the continuation of a project.

Just Working for a Salary

"I feel like I am just working for a salary," Jane suddenly asserted with some degree of frustration. "Working on this project is not very satisfying to me personally. We were partners for the first six months of the project, but once the project was started, I felt like I was just implementing written rules – the intervention was written out, the training was written out, and the data collection was written out. The training was not created by us, the database was not created by us, and the data analyses were not designed by us. So, the plan all along was for the United States to use us as subcontractors, not as partners. All I am doing is implementing this project – I have no time to do anything else."

"We did not know these are problems for you," Chris responded with some surprise. "Perhaps we should back up and make sure we share a common purpose for this project. Bob and I are doing this project to help the children, both the children currently in the project and perhaps vulnerable children in the future. We are guided by what is good for the project, and we assume that what is good for the project is good for each individual partner. And to accomplish this we try to help and support you."

George had a similar purpose. "I want to change children's lives. I get satisfaction from the project, because I see that the caregivers and the children are changing and improving. I also want to see these changes implemented in other services, and I know we need the research part of the project to make a stronger case for spreading the principles of the project to other organizations and to policymakers."

"It's unfair to say that we only do this project for the children," countered Jane. "Take data analyses, for example. Originally Bob thought the database would be housed only in the United States and the data analyses were best done in the United States."

"I admit," Bob confided, "that was naïve and inappropriate on my part. I made the assumption – without asking about it – that you all were less experienced at data analyses than we were, and that we had a full-time statistician/data analysis person on the project in the United States."

"We had to strategize and play games to get you to change your mind," Jane asserted.

"You did not have to trick us to get us to agree for you to have a complete copy of the database and to analyze data," countered Chris." When you asked for that, we immediately agreed, and Bob has been devoting entire days on each trip to discussing data analysis strategies with you and your statistician. And now you say you do not have time to conduct data analyses. We don't have a free year at the end of this project to conduct data analyses, so as much of it as possible must be done while the project is still going on."

"The only time you need to trick me," Bob said with some irritation, "is if you do not have a very good reason for what you want to do. And this happened when we were deciding which assessments we would conduct. We discussed some of the more projective and psychoanalytically oriented questionnaires you proposed, and I said we could not ask the service personnel to fill out so many questionnaires. Some of those instruments that you suggested did not have sufficient validity and psychometric information to empirically justify their relevance. We also probably could not get them by journal reviewers. We all agreed that we would not administer them. Months later we found out that you gave them anyway without telling us."

George sensed the discussion was getting tense and changed the subject, but not by much. "The article Chris has drafted for publication describing the project has Chris as the first author and Bob as the last author. The first and last authors are the most important, so it looks like a US project not a real partnership. Jane had previously written about certain changes that might be introduced in such services in her chapter before we planned them for this project. So, she should be the first author of an article describing the intervention."

"If this was in her chapter," countered Chris, "why were some of these components of the proposed intervention opposed at first? In any case, I think we need to review how this article came to be. We all made a presentation on the project at a professional meeting, and afterward the editor of the society's journal asked the entire group of us to submit an article describing the project. No one volunteered to write it. So eventually I said I would draft it for the group, and the original draft I sent to you indicated 'authors to be named.' So, it was not initially assumed that I would be the lead author, and we waited to see what contribution you folks wanted to make. You did not respond – nothing. So, we went ahead and submitted the manuscript because we did not want to lose the opportunity."

"Also," Bob added, "the international principles governing the sequence of authors on an article say that authors should be listed roughly in the order of the size and importance of their contribution. Chris led the planning of the intervention, and Jane told me after those sessions that 'you all knew what you wanted but did not know how to plan it.' She also wrote the article. Years ago, especially in the medical community, the last author was very important – it was often the Chair of the Department or Laboratory who may not have contributed anything to the project and article. But this is now specifically not allowed, so now the last author, in this case me, had the least to do with the substance of the article."

Jane moved to another concern. "I am honored and grateful to have received a grant of $1,000 from the society to help me attend its next convention in the United States, but I am disappointed in my role as a discussant on our symposium. It makes me look like I am not a participant in this project. It's another example of how the US Team takes the credit for this project."

"We understand," admitted Bob, "and this was not what we wanted either, but we did the best we could under the circumstances. Initially, we asked you if you wanted to participate in the symposium proposal, and you said you could not guarantee that you would have the money to attend. OK, so we decided to submit a symposium proposal with you as a discussant. We also submitted an application for you to receive a travel award, but we did not know how much the travel award would be and whether it would be sufficient for you to attend. We did not tell you we did this, because we did not want you to get your hopes up and then perhaps be disappointed. Maybe that was a mistake.

"Most organizations have a policy that states that each person proposing to present as part of a symposium guarantees that they will indeed attend and present if the symposium is accepted. We could not guarantee that you would attend. The role of a discussant on a symposium is somewhat different. First, the discussant is usually someone who has broad knowledge of the symposium topic and is often better known than the other presenters – it is not a low-level role. Second, discussants do not propose a specific presentation, so if they do not attend, the symposium goes on more or less as proposed. We indicated in our proposal that your attendance was contingent on getting a travel award. If not, we would get another discussant. This was our way to respect the rules and a way to get you to be part of the symposium. We hope the award is sufficient for you to attend and present. Perhaps we should have described all this to you at the time."

"But," Chris continued on another topic, "you say that you are just following the rules and written prescriptions, but both Teams decided together what the procedures would be and what each Team would do.

"You said it was a partnership at the beginning of the project, which is when we planned everything together," observed Chris. "Apparently you are used to working freely and independently, but one does not work freely and independently in a partnership – that is what a partnership means."

"We know that the Americans are always open to our suggestions," acknowledged William. "And the Americans do indeed change their mind when we make suggestions. But they always offer their own version."

"Yes," admitted Bob, "we do suggest how to do things; we have a great deal of experience doing precisely this kind of work and in a partnership manner. But we also need your knowledge and experience, and we ask for your suggestions. In big intervention projects such as this, implementation of the intervention and the assessments is a crucial component. A great intervention program poorly implemented is a poor program, and you have the major responsibility for implementing this project and assuring its quality. Implementation consists of a great deal of monitoring, meetings, making what seem like small but very important decisions, and so forth. It appears very tedious, but it is also creative. We know, we have done this sort of thing numerous times. But you need to keep us informed – you did for a while, but less often recently. We frequently do not get a response from you, as we just mentioned."

"That's fair," George nodded in agreement.

"We expect colleagues and partners to make suggestions and to criticize," Chris emphasized. "For us, only subordinates do not criticize. When you are colleagues, you need to tell us right away when you disagree or think something should be changed."

"In our country," advised William, "we don't say these things."

"Well, that's a problem," muttered Chris.

"We want to do this project," George insisted. "There is much we can accomplish together. But it is not unusual for the two sides to see things differently."

"I did not mean these comments to be complaints," Jane allowed. "Let's continue this discussion tomorrow."

"I'll begin the discussion," Chris offered the next day. The atmosphere was already a bit tense. "Frankly, we were not only shocked by what we heard yesterday, but also insulted. For example, I said I was doing this project for the children and for you, and George said nearly the same thing. You completely ignored those purposes and implied that we were motivated by self-interest in the form of authorship and imposing our ideas on you and the project."

"And I was upset," added Bob, "when you said you could not believe most of what I said, you had to trick me to allow you to analyze some of the data, and you really don't trust us.

"Partnerships are a different way of conducting interventions and evaluations. I did not know how to operate a partnership when we started the Office of Child Development at our university, but Chris had many experiences with large partnerships with community agencies and she taught me."

"Partnerships," continued Chris, "are needed when you cannot do a project on your own, but you need the experience and skills of others to do it well or to do it at all, which is the case for our project. The price one pays for obtaining those skills from someone else is that you must share control, responsibility, and benefits. It's a little like a marriage – you have to give up something, such as total control, credits, and benefits, to get something you could not have achieved alone. Yesterday, you gave us the impression that you wanted us to give you the money and consult on statistics three times a year. That is not a partnership.

"Partnerships involve divisions of labor and sharing control. When we ask for your suggestions and contributions, we frequently do not get any response from you, especially to our more recent written requests. We know it is difficult because everything is in English, but we cannot help that, and we really do not think language is the problem because you speak English quite well. It is not a partnership if you don't respond and make suggestions for changes. That lack of responsiveness also makes it appear to YOU that you are 'just working for a salary.'"

"I'm very worried about us," Chris went on. "We cannot focus on who came up with which idea; this will be very destructive to our relationships. We should have a project that we are ALL proud of. We were surprised and shocked when you said you have never trusted us from the beginning. I fear that we will leave in a few days and we will not trust each other in the future to finish the project. Without trust, it will not work. We are very disappointed and upset. We have spent too much time together to have one day destroy it all."

"We did not plan to say all that," William admitted. "It was painful, and it did not reflect the last three years very accurately. We could have handled it better. But I do not have a clear idea of how to resolve these issues."

"I am very sorry about yesterday's conversation," George admitted. "It saddened me, too. I did not sleep much last night. I was shocked at some of the words that were used. The hurtful words are our problem.

"I spend all my time trying to make life better for the children, and if someone else has another priority for this project it will be difficult for me to continue. You need to trust that my motivation is for the children."

"I don't think all the ideas for this project came from us," George continued. "I do not want to divide up which idea was whose. We do not have much experience working in a partnership and do not seem ready for shared control and giving up something to get something else.

"We need to think about how to organize our work better. Originally, I was to be the Administrative Director, but then Jane took that over. Now she feels overwhelmed with it, but I do not want that responsibility back. I like this project because it is not totally one person's responsibility. We need to focus on the project goal and not destroy it from the inside, because there is enough resistance to it from the outside. God save us from those outsiders."

After more discussion of this topic, sometimes rather heated, William declared that we should stop and move on to more practical procedural issues. That cleared the air a bit, but it did not solve the problem. Chris and Bob left for home, but we stewed about all this continually for several days. Then, we decided to write a statement of our feelings and to outline the basis for our relationship in the future. We voiced concern that our collaboration lacked mutual trust, which is an essential component of a successful collaboration.

Our colleagues emailed a response saying that they could not be simply executors of the project and that their creative and professional work had to be respected as equal partners.

Toward Understanding

As time passed, emotions mellowed a bit on both sides, and Chris and Bob tried to put themselves in the position of the international team members and tried to understand their perceptions and feelings.

First, our two countries had been somewhat antagonistic for many years, and neither government trusted each other very much. Perhaps some of this political history and stereotypes influenced perceptions and expectations of each other.

Second, there is no question that the subcontract with our international partners appears to pay for a work for hire. Indeed, our colleagues objected to the first draft of the contract and wanted a public relations document about why this project should be done and its benefits. We told them that an American contract basically states, "Here is some money, here is what you will do for that money, and here is what will happen if you don't do it." Well, given their lack of experience with such contracts, no wonder they saw this as hiring them to work for us.

Third, our international colleagues liked the partnership style of project creation because they had many of the ideas for how to change the services before we got there. They had these ideas for many years but did not have a plan to implement them, which Chris led all of us in designing together. The initial stages of implementing the intervention and the measurements took a great deal of effort and creativity on their part. But once thoroughly established, everything had to be written down and implemented as written. A project like this then does become more routine and without as much additional conceptual creativity. Since they had never done anything like this on this scale, perhaps they did not expect this. Implementing a heavy data collection effort like this project is very tedious. It is often said that such research is "95% perspiration and only 5% inspiration," and this is what that means. No wonder after the first year of creating the project they felt like they were working for a salary with little additional professional creativity. But they conducted implementation very well, and it actually took a great deal of procedural creativity on their part.

Fourth, their perception that the task of implementation was not "scholastically creative" or "professional work" may have been influenced by a bit of "academic culture shock." For decades, the discipline of psychology worldwide largely consisted of thinking, conceptualizing, and theorizing about behavior. Psychoanalysis was an example, and two of our colleagues were trained in it. But in the mid- to late 1900s, the US government invested in behavioral research, and psychology as a discipline in the United States became much more empirical, and data collection and analyses predominated. Our project was an example. So, while the creation of the intervention and deciding on measurements were part of traditional psychology to our colleagues, the management of the intervention, data collection, and data analyses were non-traditional activities for psychologists who perhaps did not perceive them as part of their professional activity.

Fifth, from their perspective, they may think they are doing all the work on this project. They don't see what we do back in the United States. It takes a great deal of work and professional creativity to write federal grant applications, to write an ethics application in the United States, to conduct the continuing administrative and financial responsibilities, to analyze all the data, and to write a book-length technical report for publication and respond to reviewers' criticisms. Yes, they see some of these products, but they don't see the day-to-day effort it takes. So, to them, they are doing most of the work and should get most of the credit.

Sixth, our colleagues had every right to be angered by the suggestion of not having the database reside with them and to have no plan or money for them to conduct data analyses. That was unthinking and done without investigating their desire to conduct data analyses. From their perspective, it was American domination and control, a professional putdown, and it contributed to their perception that they were working for the Americans.

Finally, our international colleagues were not the only ones who did not always communicate fully and promptly; sometimes we did not either. There is no substitute for prompt, honest communication.

Reconciliation

Such thinking was a start on reconciliation. But more was needed to begin to restore relationships, so Chris suggested to the group that we all have a retreat meeting at some nearby city away from our usual meeting grounds and that we consider having a mediator to help us restore relationships. We asked our colleagues if they knew of someone who was experienced at this sort of thing to mediate for us. They did not.

So, we proposed two individuals who were quite different in backgrounds and approaches. One was a clinical psychologist in Pittsburgh who directed a youth services agency and had a great deal of experience at conflict resolution and team building within service agencies. Although Chris had known this individual for many years, we have had only one project with him several years ago, so we felt he could be an independent mediator. He would take a more psychological orientation toward relationship issues, seeking to help individuals understand themselves and the others and how to work better together.

The second possible mediator was a retired career diplomat and US Foreign Service negotiator with substantial experience negotiating international arrangements. He had extensive experience with cultural differences, and he would attempt to get us to agree on good working arrangements and relationships. We had no other relationships with this professional.

Both these professionals were willing to meet with all of us for several days. While they were largely independent of Chris and Bob, they were both Americans. So, our international colleagues understandably decided against a mediator but did vote for a retreat meeting among ourselves.

The group rented a van and held the retreat in a town a few hours away from our base. The various issues were discussed, and the tone was more sensitive and accommodating than before. Also, we decided that at the end of each future visit, the group would talk about how each person perceived the process as unfolding. Was the tone of the discussion good? What did each person like best about the meetings? What did not go so well for each person? Were all disagreements voiced in the discussion? Did we all try to understand, respect, and accommodate to disagreements? Were there still lingering concerns? What could we do better or change at the next meeting?

Ultimately, the two teams managed to work together well enough to finish the original project and to plan and execute a subsequent project.

Conclusion

Collegial planning and relationship building are crucial first steps, but relationships need to be nurtured throughout the project. From a practical standpoint, perhaps true equality among partners is a myth, and that reality needs to be recognized. It helps to periodically step back from the collaborative activity and ask the group how they feel the process is going and how it could be improved, because the process is crucial to the quality of the project. Further, a little empathy can be beneficial – how do our partners perceive various aspects of the process? Some components that are routine for you (e.g., subcontracts) may be perceived very differently by your colleagues. Of course, collaborations are composed of individuals who bring their own personalities and perceptions to the process and who perceive situations and responsibilities differently at the beginning and over the course of the project. Do not necessarily assume that if things are going well on the surface that they are going well beneath the surface. Therefore, spend some time at each meeting monitoring and discussing the process, not just the project.

Part IV

Conclusions

Although much of this book has presented guidance, warnings, and illustrations of problems, if not disasters, there are many benefits of conducting international collaborations, which we describe in the next chapter.

17 Benefits of International Collaborations

There are a variety of possible benefits to your country and the country you worked in. We list below some of the contributions we think our international work has made as examples of the potential benefit of this kind of activity.

Benefits to International Collaborators and Countries

The most direct benefit of an international project is likely to be its success at achieving its goals. In our case, it was demonstrating that changes made in the structure of agencies and the way caregivers interacted with the children produced substantial improvements in children's development. In one country, our international colleagues convinced the national legislature to pass a resolution that all such agencies serving vulnerable children in the country should be structured in a manner similar to our family-like intervention.

This resolution was an unfunded mandate and did not require special behavioral training for caregivers. Nevertheless, many agency directors have visited our intervention agency and have been motivated to improve their services. Our international colleagues have given numerous presentations and training sessions to administrators and directors of agencies and helped them modify their services. They are also providing training to organizations that support foster care and adoption services, which are now expanding in their country. Moreover, one of our international colleagues was sent to North Korea under UNICEF auspices to make suggestions to them on how their services for vulnerable children could be improved. And we have worked with a foundation in another country to help them improve their services and to prepare professionals to support adoptive and foster parents.

We believe the results of our projects provide a well-documented illustration for academics and service professionals in their country and elsewhere of the importance of children's early behavioral experiences on their physical as well as cognitive and behavioral development. One of our projects is the best demonstration that positive early caregiver–child behavioral interactions improve children's height and weight, and it was conducted in a country that did not embrace behavioral pediatrics at the time. Eventually, we suspect, word will get around that caregiver–child interactions and relationships are important contributors to children's medical status.

A corollary contribution, we hope, is that agency directors and pediatricians will revise their belief that nothing can be done to improve the functioning and well-being of children with disabilities. One of our projects showed that most children with disabilities improved in their functioning and cognitive/behavioral scores as a result of our intervention. We hope this result gets communicated more widely in that country and that behavioral interventions for children with disabilities become more commonplace there.

We provided research support for family care. Our projects are sometimes perceived as evidence that agencies should be improved, perhaps instead of promoting family placements. Instead, one of our projects can be perceived as the best evidence available that the characteristics of good families indeed promote appropriateommunicated more widely in that country and that behavioral interventions for children with disabilities become more commonplace there.

We provided research support for family care. Our projects are sometimes perceived as evidence that agencies should be improved, perhaps instead of promoting family placements. Instead, one of our projects can be perceived as the best evidence available that the characteristics of good families indeed promote appropriate development in vulnerable children. Our intervention took each major characteristic of a good family and made the agency structurally more family-like and the care provided to the children more parent-like. These characteristics of a good family, even when implemented in the context of a service agency, produced vastly improved child development. Our project provides some of the best available evidence that placement of children in good families is likely to produce good child development. Therefore, it supports, not distracts from, family-based child welfare systems.

We left an enduring legacy. In at least one of our projects the main intervention agency was able to continue the new form of care on their own budget for many years after our project ended, and research showed children continued to benefit. In addition, we developed a written curriculum that was translated into the local language to train professionals and caregivers to interact with children in a more engaged, sensitive, and responsive manner. This material could be used indefinitely to train new personnel and those working in other contexts.

We believe our international colleagues and their students benefited professionally and personally. Two of our international colleagues are on the faculty at their local university and were authors on numerous publications over the last 15 years. This is quite noticeable as their university moves toward obtaining more grants and publishing more academic papers.

We gave talks to faculty and students on the university–community engagement center we co-directed at the University of Pittsburgh as well as on topics relating to conducting applied research projects. Also, one project provided data for many students, including one of our international colleagues, to obtain advanced degrees and publish papers. One young faculty member came to the United States for several weeks to work with us and our colleagues to sharpen her statistical and academic writing skills, and she eventually published the paper that was the focus of her visit.

Our project provided data and opportunities to our international colleagues and their students. Our colleagues and some of their students have also presented some of our project's findings and their related work at several international conferences over the last decade. These experiences gave them the opportunity to hear about other research from around the world and to meet and expand relationships with scholars from many countries. We also funded and organized an invited conference of many of the world's most prominent scholars on care for vulnerable children, and one of our international colleagues participated.

The project was a major administrative learning experience. Implementing some of our projects represented a relatively new experience for our international colleagues in every aspect of conducting a very large, comprehensive, empirical project, including the grant application process, ethics reviews, implementation of the intervention, financial and accounting practices, statistical analyses, and publishing practices.

We provided independent documentation that the approach of other professionals and funders was effective at improving children's development. These funders and service professionals generally had limited knowledge of evaluation practices, and our work provided credible evidence that their programs were effective.

In all the countries, we gave lectures and workshops on a variety of topics. Chris, a specialist on how to handle, position, and engage children with disabilities, frequently gave presentations on these subjects in each country to administrators, professionals, and caregivers. These countries typically lacked nearly any training in these matters and tacitly or overtly believed nothing could be done for these children. Chris would emphasize that these children typically are more capable than they appear, and because they often cannot function in the usual ways, caregivers need to keep "trying another way to reach and encourage these children."

In one country, our project was to train their psychologists, social workers, and other professionals on modern methods for services for vulnerable children and to support parents who may foster or adopt these children. This included how to promote adult–child relationships, how to improve children's mental development, how to help children cope with trauma, and how to deal with children's problem behavior. Further, our training introduced "family therapy," in which child and parent are treated together, which was virtually unknown in some countries. We also provided a host of concrete actions parents and counselors could take to deal with behavior problems that are common among vulnerable children. This workshop was written and translated into the local language so professionals could train others throughout the country, and they continue to conduct those trainings.

Benefits to Our Own Country

We answered the focal scientific question. In one project, the scientific purpose was to determine the extent early caregiver–child interactions and relationships were necessary for typical development of infants and toddlers apart from medical care and nutrition. While this has been assumed for decades and much research shows that these early experiences influence children's development, it was rare that children resided in environments that were so devoid of such adult–child engagement and where the development of children was so delayed (on average below more than nine out of ten parent-reared children). Our project showed that a behavioral intervention alone, without changes in nutrition, medical care, sanitation, and safety, could improve children's physical growth and cognitive/behavioral development. Moreover, it did so for children with and without disabilities, and the improvements were greater the longer children were in the intervention condition. Those children who remained in the intervention for two or more years approached the developmental levels of family-reared children. From a scientific standpoint, the research question was clearly answered.

We answered a second scientific question. In a follow-up project, we studied the development of children who spent their early months in our improved and comparison agencies and were then transitioned to the United States and domestic families. Generally, most such children show remarkable catch-up improvement once they enter families, but a higher-than-expected percentage of such children show lingering cognitive and behavior problems. Children from our improved agency displayed better attachment to their new parents and fewer behavioral problems than children who experienced the care-as-usual agencies. Therefore, our intervention had lasting benefits.

We trained numerous US students. In the course of nearly two decades of international work, we trained seven graduate students through their doctorates, several of whom published numerous papers using our data, and we took several students and staff members with us to our project sites in three foreign countries.

We tried to implement the results of our intervention in US childcare. Of course, the children in US childcare and children in the agencies we worked in abroad are quite different. But we explored implementing some of the principles of our intervention in childcare environments in the United States. Would children in childcare be improved if they were placed in mixed-age rather than homogeneously-aged groups? Would they do better if they stayed with the same teacher and perhaps some of the same peers for several years (called "looping") instead of "graduating" to new groups each year? Would very young children improve more broadly if their care were more socially-emotionally focused rather than an emphasis on academic-type learning? Each of these questions has received some attention, but they have never been all put together in a single experimental early learning or childcare facility.

We explored creating a new kind of early care based on these principles. We quickly ran into regulations on group size, children:caregiver ratios, and even square footage of space per child, all of which were keyed to the specific ages of the children in the group. Mixing ages cut across these policy lines. We would need waivers of these regulations to try it. That was possible, we were told by our state's administrator in charge, but not politically wise at the time, because legislators opposed to funding childcare were looking for any excuse to cut those budgets. Now, the political climate for early care and education has improved, and it might be time to try such an innovative facility.

General Potential Benefits of International Collaboration

It may broaden the world's knowledge base. Many international research and service projects are initiated because the country hosting the project provides a group of people or circumstances that are not available in a Western nation. For example, as observed in Chapter 1, most behavioral research and service knowledge is based upon a few specific populations available in the West. International collaborations provide the opportunity to discover whether the same principles also apply to somewhat different groups of people. Also, because of the different backgrounds of people and circumstances in which they live, new factors may be discovered that influence major behavioral phenomena and the effectiveness of services.

It can spread technical skills and knowledge to other countries. Western nations have developed superb training programs, advanced treatments of various conditions, and technical skills in conducting behavioral research and services. These assets can be offered to other countries to blend with their local circumstances. The science of program evaluation, for example, is often lacking in some countries in which major new programs are being created, and Western behavioral researchers could contribute to training individuals in such nations on these skills. Similarly, as we saw, services for children with disabilities in some countries are quite antiquated and potentially ineffective or even harmful. More modern techniques would be helpful.

It can provide an opportunity to view ourselves from a different perspective. Sometimes behaviors and customs that we consider commonplace are perceived quite differently by people in other countries. We made several mistakes and encountered numerous problems, some of which were the result of us not perceiving our own actions the way others saw them. Sometimes it helps to see ourselves as others see us and why they do so.

It provides an opportunity to be good ambassadors. International working groups provide an opportunity for people to have face-to-face, concrete experiences with people from other countries. Sometimes our knowledge of another country is based upon the behavior of their government, but the people of that country can be much different than this political persona.

We tried to work as collaboratively as possible under the circumstances, and to be respectful and gracious with our local colleagues. We hope we were successful most of the time and gave many of our colleagues a good impression of Americans – diplomacy one or two people and projects at a time.

Index